Artificial Intelligence for Robotics

Build intelligent robots using ROS 2, Python, OpenCV, and AI/ML techniques for real-world tasks

Francis X. Govers III

‹packt›

Artificial Intelligence for Robotics

Group Product Manager: Preet Ahuja

Publishing Product Manager: Surbhi Suman

Book Project Manager: Ashwini Gowda

Senior Editors: Shruti Menon and Adrija Mitra

Technical Editor: Rajat Sharma

Copy Editor: Safis Editing

Proofreader: Safis Editing

Indexer: Pratik Shirodkar

Production Designer: Vijay Kamble

DevRel Marketing Coordinator: Rohan Dobhal

Senior DevRel Marketing Coordinator: Linda Pearlson

First published: August 2018

Second edition: March 2024

Production reference: 1290224

Published by Packt Publishing Ltd.

Grosvenor House

11 St Paul's Square

Birmingham

B3 1RB, UK.

ISBN 978-1-80512-959-2

www.packtpub.com

In loving memory of my grandmother, Sally (Mabel) Govers, who nurtured my love of science and learning from my earliest days.

Thanks to my long-time robotics mentor, Dr. Bob Finkelstein.
Thanks, Bob, for all your faith and encouragement.

To my wife, Carol – thanks for your patience. To my children, Jessica and Corbin, their spouses, Peter and Amy, and my grandchildren, William, Oliver, Amelia, Henry, and Rowan. This book is dedicated to you all; thanks for the inspiration.

Foreword

It is indeed a pleasure and an honor to prepare this foreword. *Francis X. Govers III* has four decades of experience in designing autonomous vehicles – self-driving cars, airships, unmanned aircraft, and robots. The content in this book reflects his expertise, knowledge, and authority on the subject of **artificial intelligence** (**AI**) for robotics. Francis has carefully crafted the chapters keeping in mind a reader who is eager to learn the practical aspects of the subject, taking them to a higher level where they feel they learned the topic to the extent that they can continue learning on their own. Francis showed his wizardry in teaching the art of AI by organizing the book into three learning modules and breaking down each module into three chapters. The final two chapters capture the summary and assessment of the knowledge gained.

This book is for professionals or hobbyists who have theoretical knowledge and are interested in applying it by learning how to program a robot to perform intelligent tasks autonomously. Each chapter in this book begins with a concept within the realm of AI, a programmable task, and a means to implement the task on a robot with illustration. While each chapter is sufficiently independent, they are designed to build from the previous ones with increasing depth and complexity. This is the magic thread that runs through the book consistently. Each chapter makes the learner accomplished, giving the satisfaction of learning. Learners can go through this book at their own comfortable pace, one chapter at a time, making progress gradually.

This takes the learners through the world of autonomy gradually. Francis guides the learners and makes them confident in each topic, yet curious to learn more. From setting up a robot for the first time to making the robot express emotions, Francis makes learning a fun activity! The subject of autonomy is complex, yet this book makes it easier to understand through illustrated examples and sample code segments. I am confident that this book will serve as a hands-on guide on intelligent robotics for college freshmen and professionals alike to fulfill the intent of the author.

Dr. Kamesh Namuduri

Professor, Department of Electrical Engineering

University of North Texas

Contributors

About the author

Francis X. Govers III is an Associate Technical Fellow for Autonomy at Bell Textron, and chairman of the Textron Autonomy Council. He is the designer of over 30 unmanned vehicles and robots for land, sea, air, and space, including RAMSEE, the autonomous security guard robot. Francis helped lead the design of the International Space Station, the F-35 JSF Fighter, the US Army Future Combat Systems, and telemetry systems for NASCAR and IndyCar. He is an engineer, pilot, author, musician, artist, and maker. He received five outstanding achievement awards from NASA and recognition from Scientific American for World Changing Ideas. He has a Master of Science degree from Brandeis University and is a veteran of the US Air Force.

Thanks to my teammates at Bell Textron, especially Jason Hurst, Keith Stanney, Lee Anderson, Grant Bristow, Matt Holvey, Srushti Desai, Jeff Holcomb, and Rohn Olsen, and all the support from across the Textron organization. I also thank my academic partners, Dr. Kamesh Namuduri at the University of North Texas, and Dr. Kamesh Subbarao at the University of Texas at Arlington.

About the reviewers

Jugesh Sundram is a robotics engineer working in the automotive and mobility sector on industrial problems that require computer vision-based AI solutions. He obtained his Master's in mechanical engineering from Florida Institute of Technology, USA. He has worked across the globe, spanning the US, India, Singapore, and now Belgium. He focuses on researching, designing, developing, and applying simultaneous localization and mapping algorithms on mobile robots. He has contributed to research papers in various top robotics conferences and holds patents. A father of two, he enjoys his time off spending time with family, producing music, traveling, and exploring vegetarian cuisine across the world.

I attribute my joy in reviewing to the curiosity and knowledge imparted to me during my academic journey. For that, I thank all my teachers, professors, and colleagues who have mentored me along the way and led me to where I am today. I am grateful to my wonderful wife, kids, and parents, who gave meaning to the word "life" in work-life balance.

Karthikeyan Yuvaraj has over 10 years of experience in developing and deploying complex robotics and perception systems. His experience lies in developing software solutions for vision-based robotic manipulation. He has also worked in major industry names such as Alphabet, Honeywell, and DARPA. Currently, he is a senior camera software engineer at HP Inc., where he develops camera systems for state-of-the-art video conferencing technologies. He is also a fellow of the Institution of Engineering and Technology and a senior member of the Institute of Electrical and Electronics Engineers.

Table of Contents

3

Conceptualizing the Practical Robot Design Process   53

Part 2: Adding Perception, Learning, and Interaction to Robotics

4

Recognizing Objects Using Neural Networks and Supervised Learning   87

5

Picking Up and Putting Away Toys using Reinforcement Learning and Genetic Algorithms   111

6

Teaching a Robot to Listen   147

Part 3: Advanced Concepts – Navigation, Manipulation, Emotions, and More

7

Teaching the Robot to Navigate and Avoid Stairs   173

8

Putting Things Away   201

9

Giving the Robot an Artificial Personality   237

10

Conclusions and Reflections   269

Answers   281

Appendix   299

Preface

The definition of a robot is a machine that can do human-like tasks. In order to perform these tasks, a robot must be able to see, understand, and interact with the environment. AI is the quickest way to recognize objects and navigate. This book empowers you with the essential skills to efficiently operate your robots using AI techniques such as **Convolutional Neural Networks** (**CNNs**), computer vision, object recognition, genetic algorithms, and reinforcement learning.

So, who is writing this book? As you can see from my biography, I've been doing this for over 40 years, starting as a 12-year-old kid doing science fair projects, then as an Air Force enlisted person, an entry level engineer at NASA, and so on. I started doing AI in 1992 and specialized in machine decision making. Today, I design complete flying autonomous vehicles that weigh tens of thousands of pounds. I'm pleased to be putting some of that experience down on paper to share with you.

Why did I write this book? I felt that there was a gap in the available literature for someone coming up in the robotics and autonomy world who needs to move in capability from the hobbyist to the beginnings of industrial and commercial robotics. In doing this, I want to remove as many of what I perceive as barriers between you and what you want to accomplish as a next-level robotics explorer. I left out the equations, the strange jargon, and the mystery and replaced them with straightforward explanations about how to get what you want out of your robot. It is important to remember that my robot, Albert, is just a tool in this process. The objective of the book is not to design a singular robot, but rather to teach a set of skills that I think you need. The real question is where you go from here. Use this text as a springboard to continue to explore, experiment, and continue your education in robotics. From here you can read the works of the people who inspired and/or mentored me: Dr. Robin Murphy, Sebastian Thrun, Dr. Rodney Brooks, Bob Zubrin, Dr. Robert L. Forward, Isaac Asimov, Arthur C. Clarke, and many others.

Who this book is for

This book is for robotics engineers and enthusiasts who have already begun their journey of learning about robotics and wish to progress to a more advanced stage of capability by applying AI. It will be a useful reference for students and researchers looking for a practical guide to solving specific problems or approaching a difficult robot design. Basic programming skills in Python, a familiarity with electronics and wiring, and the ability to use a Linux-based **command line interface** (**CLI**) will be useful while reading this book.

What this book covers

Chapter 1, The Foundation of Robotics and Artificial Intelligence, explains what we will cover in this book, standard robot parts, concepts of control, computing in real time, and the **Observe, Orient, Decide, Act (OODA)** concept for how robots make decisions.

Chapter 2, Setting Up Your Robot, introduces you to motors, control systems, how to divide the robot problem into parts using the Subsumption Architecture, and the **Robotics Operating System 2 (ROS 2)**.

Chapter 3, Conceptualizing the Practical Robot Design Process, describes systems engineering techniques for robot design, use cases, and storyboards.

Chapter 4, Recognizing Objects Using Neural Networks and Supervised Learning, explains how you can use CNNs to train object recognition and segment objects from the background.

Chapter 5, Picking Up and Putting Away Toys using Reinforcement Learning and Genetic Algorithms, covers Q-learning and genetic algorithms, which are used to teach the robot arm to move efficiently.

Chapter 6, Teaching the Robot to Listen, shows you how you can add a digital assistant to your robot and create some custom controls for it, including telling knock-knock jokes.

Chapter 7, Teaching the Robot to Navigate and Avoid Stairs, outlines how you can use another CNN to teach the robot to navigate around a house and avoid obstacles.

Chapter 8, Putting Things Away, describes how we can finish the robot's tasks, and how to find the toybox.

Chapter 9, Giving the Robot an Artificial Personality, explains the concepts and theory around simulating a personality in a robot for added interactions.

Chapter 10, Conclusions and Reflections, talks about robotics as a career, based on the author's own 40-year career in robot design.

To get the most out of this book

You should have a good grasp of programming with Python version 3. We use ROS 2 as the control architecture for the robot. Packt Publishing has several excellent books that explain how to use ROS 2 if you need more detailed instruction. Programming skills in Python, familiarity with electronics, wiring, and single-board computers, the ability to use a Linux-based CLI, and knowledge of AI/ML concepts will be required while using this book. If you want to follow the robot's construction, then basic hand tools (screwdrivers, wrenches, Allen keys, and a soldering iron) will be required. All the other installation instructions are provided in the appropriate chapters as the book progresses.

Download the example code files

You can download the example code files for this book from GitHub at `https://github.com/PacktPublishing/Artificial-Intelligence-for-Robotics-2e`. If there's an update to the code, it will be updated in the GitHub repository.

We also have other code bundles from our rich catalog of books and videos available at `https://github.com/PacktPublishing/`. Check them out!

Conventions used

There are a number of text conventions used throughout this book.

`Code in text`: Indicates code words in text, database table names, folder names, filenames, file extensions, pathnames, dummy URLs, user input, and Twitter handles. Here is an example: "We create a short Python program called `downloadDataset.py`."

A block of code is set as follows:

```
from roboflow import Roboflow
rf = Roboflow(api_key="*****************")
project = rf.workspace("toys").project("toydetector")
dataset = project.version(1).download("yolov8")
```

Any command-line input or output is written as follows:

```
cd ~/ros2_ws/src
ros2 pkg create -build-type ament-cmake ros_xarm
colcon build
```

Bold: Indicates a new term, an important word, or words that you see onscreen. For instance, words in menus or dialog boxes appear in **bold**. Here is an example: "Use the **Generate** tab on Roboflow, then click on **Add Augmentation Step** to select the type of operation that will affect our images."

> Tips or important notes
> Appear like this.

Get in touch

Feedback from our readers is always welcome.

General feedback: If you have questions about any aspect of this book, email us at `customercare@packtpub.com` and mention the book title in the subject of your message.

Errata: Although we have taken every care to ensure the accuracy of our content, mistakes do happen. If you have found a mistake in this book, we would be grateful if you would report this to us. Please visit `www.packtpub.com/support/errata` and fill in the form.

Piracy: If you come across any illegal copies of our works in any form on the internet, we would be grateful if you would provide us with the location address or website name. Please contact us at `copyright@packt.com` with a link to the material.

If you are interested in becoming an author: If there is a topic that you have expertise in and you are interested in either writing or contributing to a book, please visit `authors.packtpub.com`.

Share Your Thoughts

Once you've read *Artificial Intelligence for Robotics*, we'd love to hear your thoughts! Scan the QR code below to go straight to the Amazon review page for this book and share your feedback.

`https://packt.link/r/1805129597`

Your review is important to us and the tech community and will help us make sure we're delivering excellent quality content.

Download a free PDF copy of this book

Thanks for purchasing this book!

Do you like to read on the go but are unable to carry your print books everywhere?

Is your eBook purchase not compatible with the device of your choice?

Don't worry, now with every Packt book you get a DRM-free PDF version of that book at no cost.

Read anywhere, any place, on any device. Search, copy, and paste code from your favorite technical books directly into your application.

The perks don't stop there, you can get exclusive access to discounts, newsletters, and great free content in your inbox daily

Follow these simple steps to get the benefits:

1. Scan the QR code or visit the link below

https://packt.link/free-ebook/9781805129592

2. Submit your proof of purchase
3. That's it! We'll send your free PDF and other benefits to your email directly

Part 1:
Building Blocks for Robotics and Artificial Intelligence

The first section of this book begins with the foundations of robotics and **Artificial Intelligence (AI)**, covering what AI is and how it is used. Then we start to define our robot systems and talk about control. In the second chapter, we look at robot anatomy, the parts of a robot, and discuss autonomy principles and the Subsumption Architecture concept. You'll get a primer on the **Robotics Operating System (ROS)** and our single-board supercomputer. Finally, we show a systematic process for robot design using systems engineering principles and storyboards.

This part has the following chapters:

- *Chapter 1, The Foundation of Robotics and Artificial Intelligence*
- *Chapter 2, Setting Up Your Robot*
- *Chapter 3, Conceptualizing the Practical Robot Design Process*

1

The Foundation of Robotics and Artificial Intelligence

In this book, I invite you to go on a journey with me to discover how to add **Artificial Intelligence (AI)** to a mobile robot. The basic difference between what I will call an **AI robot** and a more **regular robot** is the ability of the robot and its software to make decisions and to learn and adapt to its environment based on data from its sensors. To be a bit more specific, we are leaving the world of pre-coded robot design behind. Instead of programming all of the robot's behaviors in advance, the robot, or more correctly, the robot software, will learn from examples we provide, or from interacting with the outside world. The robot software will not control its behavior as much as the data that we use to train the AI system will.

The AI robot will use its learning process to make predictions about the environment and how to achieve goals, and then use those predictions to create behavior. We will be trying out several forms of AI on our journey, including supervised and unsupervised learning, reinforcement learning, neural networks, and genetic algorithms. We will create a digital robot assistant that can talk and understand commands (and tell jokes), and we will create an **Artificial Personality (AP)** for our robot. We will learn how to teach our robot to navigate without a map, grasp objects by trial and error, and see in three dimensions.

In this chapter, we will cover the following key topics:

- The basic principles of robotics and AI
- What is AI and autonomy (and what is it not)?
- Are recent developments in AI anything new?
- What is a robot?
- Introducing our sample problem
- When do you need AI for your robot?
- Introducing the robot and our development environment

Technical requirements

The technical requirements for completing the tasks in this chapter are described in the *Preface* at the beginning of this book.

All of the code for this book is available on the GitHub repository, available at `https://github.com/PacktPublishing/Artificial-Intelligence-for-Robotics-2e/`.

The basic principle of robotics and AI

AI applied to **robotics development** requires a different set of skills from you, the robot designer or developer. You may have made robots before. You probably have a quadcopter or a 3D printer (which is, in fact, a robot). The familiar world of **Proportional-Integral-Derivative (PID)** controllers, sensor loops, and state machines is augmented by **Artificial Neural Networks (ANNs)**, expert systems, genetic algorithms, and searching path planners. We want a robot that does not just react to its environment as a reflex action but has goals and intent – and can learn and adapt to the environment and is taught or trained rather than programmed. Some of the problems we can solve this way would be difficult, intractable, or impossible otherwise.

What we are going to do in this book is introduce a problem – picking up toys in a playroom – that we will use as our example throughout the book as we learn a series of techniques for applying AI to our robot. It is important to understand that, in this book, the journey is far more important than the destination. At the end of the book, you should have gained some important skills with broad applicability, not just learned how to pick up toys.

What we are going to do is first provide some tools and background to match the infrastructure that was used to develop the examples in the book. This is both to provide an even playing field and to not assume any practical knowledge on your part. To execute some of the advanced neural networks that we are going to build, we will use the GPUs in the Jetson.

In the rest of this chapter, we will discuss some basics about robotics and AI, and then proceed to develop two important tools that we will use in all of the examples in the rest of the book. We will introduce the concept of soft real-time control, and then provide a framework, or model, for creating autonomy for our robot called the **Observe-Orient-Decide-Act (OODA)** loop.

What is AI and autonomy (and what is it not)?

What would be the definition of AI? In general, it means a machine that exhibits some characteristics of intelligence – thinking, reasoning, planning, learning, and adapting. It can also mean a software program that can simulate thinking or reasoning. Let's try some examples: a robot that avoids obstacles by simple rules (if the obstacle is to the right, go left) is not AI. A program that learns, by example, to recognize a cat in a video is AI. A robot arm that is operated by a joystick does not use AI, but a robot arm that adapts to different objects in order to pick them up is an application of AI.

There are two defining characteristics of AI robots that you must be aware of. First of all, AI robots are primarily **trained** to perform tasks, by providing examples, rather than being programmed step by step. For example, we will teach the robot's software to recognize toys – things we want it to pick up – by training a neural network with examples of what toys look like. We will provide a training set of pictures with the toys in the images. We will specifically annotate what parts of the images are toys, and the robot will learn from that. Then we will test the robot to see that it learned what we wanted it to, somewhat like a teacher would test a student. The second characteristic is **emergent behavior**, in which the robot exhibits evolving actions that were not explicitly programmed into it. We provide the robot with controlling software that is inherently non-linear and self-organizing. The robot may suddenly exhibit some bizarre or unusual reaction to an event or situation that might appear to be odd, quirky, or even emotional. I worked with a self-driving car that we swore had delicate sensibilities and moved very daintily, earning it the nickname *Ferdinand*, after the sensitive, flower-loving bull from a cartoon, which was strange in a nine-ton truck that appeared to like plants. These behaviors are just caused by interactions of the various software components and control algorithms and do not represent anything more than that.

One concept you will hear in AI circles is the **Turing test**. The Turing test was proposed by Alan Turing in 1950, in a paper entitled *Computing Machinery and Intelligence*. He postulated that a human interrogator would question a hidden, unseen AI system, along with another human. If the human posing the questions was unable to tell which person was the computer and which was the human, then that AI computer would pass the test. This test supposes that the AI would be fully capable of listening to a conversation, understanding the content, and giving the same sort of answers a person would. Current **AI chatbots** can easily pass the Turing test and you may have interacted several times this week with AI on the phone without realizing it.

One group from the **Association for the Advancement of Artificial Intelligence (AAAI)** proposed that a more suitable test for AI might be the assembly of flatpack furniture – using the supplied instructions. However, to date, no robot has passed this test.

Our objective in this book is not to pass the Turing test, but rather to take some novel approaches to solving problems using techniques in machine learning, planning, goal seeking, pattern recognition, grouping, and clustering. Many of these problems would be very difficult to solve any other way. AI software that could pass the Turing test would be an example of **general AI**, or a full, working intelligent artificial brain, and, just like you, general AI does not need to be specifically trained to solve any particular problem. To date, general AI has not been created, but what we do have is **narrow AI** or software that simulates thinking in a very narrow application, such as recognizing objects, or picking good stocks to buy.

While we are *not* building general AI in this book, that means we are not going to be worried about our creations developing a mind of their own or getting out of control. That comes from the realm of science fiction and bad movies, rather than the reality of computers today. I am firmly of the mind that anyone preaching about the *evils* of AI or predicting that robots will take over the world has likely not seen the dismal state of AI research in terms of solving general problems or creating something resembling actual intelligence.

Are recent developments in AI anything new?

What has been is what will be, and what has been done is what will be done, and there is nothing new under the sun – Ecclesiastes 1:9, King James Bible

The modern practice of AI is not new. Most of these techniques were developed in the 1960s and 1970s and fell out of favor because the computing machinery of the day was insufficient for the complexity of software or the number of calculations required. They only waited for computers to get bigger and for another very significant event – the invention of the **internet**. In previous decades, if you needed 10,000 digitized pictures of cats to compile a database to train a neural network, the task would be almost impossible – you could take a lot of cat pictures, or scan images from books. Today, a Google search for cat pictures returns 126,000,000 results in 0.44 seconds. Finding cat pictures, or anything else, is just a search away, and you have your training set for your neural network – unless you need to train on a very specific set of objects that don't happen to be on the internet, as we will see in this book, in which case we will once again be taking a lot of pictures with another modern aid not found in the sixties, a digital camera. The happy combination of very fast computers, cheap, plentiful storage, and access to almost unlimited data of every sort has produced a renaissance in AI.

Another modern development has occurred on the other end of the computer spectrum. While anyone can now have what we would have called a supercomputer back in 2000 on their desk at home, the development of the smartphone has driven a whole series of innovations that are just being felt in technology. Your wonder of a smartphone has accelerometers and gyroscopes made of tiny silicon chips called **Micro-Electromechanical Systems** (**MEMS**). It also has a high-resolution but very small digital camera and a multi-core computer processor that takes very little power to run. It also contains (probably) three radios – a Wi-Fi wireless network, a cellular phone, and a Bluetooth transceiver. As good as these parts are at making your iPhone fun to use, they have also found their way into parts available for robots. That is fun for us because what used to be only available for research labs and universities is now for sale to individual users. If you happen to have a university or research lab or work for a technology company with multi-million-dollar development budgets, you will also learn something from this book, and find tools and ideas that hopefully will inspire your robotics creations or power new products with exciting capabilities.

Now that you're familiar with the concept of AI for robotics, let's look at what a robot actually is.

What is a robot?

The word **robot** entered the modern language from the play *R.U.R* by the Czech author Karel Capek, which was published back in 1920. *Roboti* is a Czech word meaning *forced servitude*. In the play, an industrialist learns how to build artificial people – not mechanical, metal men, but made of flesh and blood, and coming from a factory fully grown. The English translation of the name *R.U.R* as *Rossum's Universal Robots* introduced the word *robot* to the world.

For the purposes of this book, a robot is a machine that is capable of sensing and reacting to its environment, and that has some human- or animal-like function. We generally think of a robot as an automated, self-directing mobile machine that can interact with the environment. That is to say, a robot has a **physical form** and exhibits some form of **autonomy**, or the ability to make decisions for itself based on observation of the external environment.

Next, let's discuss the problem we will be trying to solve in this book.

Our sample problem – clean up this room!

In the course of this book, we will be using a single problem set that I feel most people can relate to easily, while still representing a real challenge for the most seasoned roboticist. We will be using AI and robotics techniques to pick up toys in my house after my grandchildren have visited. That sound you just heard was the gasp from the professional robotics engineers and researchers in the audience – this is a tough problem. Why is this a tough problem, and why is it ideal for this book?

Let's discuss the problem and break it down a bit. Later, in *Chapter 2*, we will do a full task analysis, learn how to write use cases, and create storyboards to develop our approach, but we can start here with some general observations.

Robotics designers first start with the environment – where does the robot work? We divide environments into two categories: structured and unstructured. A structured environment, like the playing field for a FIRST robotics competition (a contest for robots built by high school students in the US, where all of the playing field is known in advance), an assembly line, or a lab bench, has everything in an organized space. You might have heard the saying *"A place for everything and everything in its place"* – that is a **structured environment**. Another way to think about it is that we know in advance where everything is or is going to be. We know what color things are, where they are placed in space, and what shape they are. A name for this type of information is *a priori* knowledge – things we know in advance. Having advanced knowledge of the environment in robotics is sometimes absolutely essential. Assembly line robots expect parts to arrive in an exact position and orientation to be grasped and placed into position. In other words, we have arranged the world to suit the robot.

In the world of my house, this is simply not an option. If I could get my grandchildren to put their toys in exactly the same spot each time, then we would not need a robot for this task. We have a set of objects that are fairly fixed – we only have so many toys for them to play with. We occasionally add things or lose toys, or something falls down the stairs, but the toys are elements of a set of fixed objects. What they are not is positioned or oriented in any particular manner – they are just where they were left when the kids finished playing with them and went home. We also have a fixed set of furniture, but some parts move – the footstool or chairs can be moved around. This is an **unstructured environment**, where the robot and the software have to adapt, not the toys or furniture.

The problem is to have the robot drive around the room and pick up toys. Here are some objectives for this task:

- We want the user to **interact** with the robot by **talking** to it. We want the robot to understand what we want it to do, which is to say, what our intent is for the commands we are giving it.

- Once commanded to start, the robot will have to **identify an object** as being a toy or not being a toy. We only want to pick up toys.

- The robot must **avoid hazards**, the most important being the stairs going down to the first floor. Robots have a particular problem with negative obstacles (dropoffs, curbs, cliffs, stairs, etc.), and that is exactly what we have here.

- Once the robot finds a toy, it has to determine how to **pick the toy up** with its robot arm. Can it grasp the object directly, or must it scoop the item up, or push it along? We expect that the robot will try different ways to pick up toys and may need several trial-and-error attempts.

- Once the toy is picked up by the robot arm, the robot needs to **carry the toy** to a toy box. The robot must recognize the toy box in the room, remember where it is for repeat trips, and then position itself to place the toy in the box. Again, more than one attempt may be required.

- After the toy is dropped off, the robot returns to **patrolling the room** looking for more toys. At some point, hopefully, all of the toys will be retrieved. It may have to ask us, the human, whether the room is acceptable, or whether it needs to continue cleaning.

What will we learn from this problem? We will be using this backdrop to examine a variety of AI techniques and tools. The purpose of the book is to teach you how to develop AI solutions with robots. It is the process and the approach that is the critical information here, not the problem and not the robot I developed for the book. We will be demonstrating techniques for making a moving machine that can learn and adapt to its environment. I would expect that you will pick and choose which chapters to read and in which order, according to your interests and your needs, and as such, each of the chapters will be standalone lessons.

The first three chapters are foundation material that supports the rest of the book by setting up the problem and providing a firm framework to attach the rest of the material.

The basics of robotics

Not all of the chapters or topics in this book are considered *classical* AI approaches, but they do represent different ways of approaching machine learning and decision-making problems. We will be exploring together the following topics:

- **Control theory and timing**: We will build a firm foundation for robot control by understanding control theory and timing. We will be using a soft real-time control scheme with what I call a **frame-based control loop**. This technique has a fancy technical name – **rate monotonic scheduling** – but I think you will find the concept intuitive and easy to understand.

- **OODA loop**: At the most basic level, AI is a way for the robot to make decisions about its actions. We will introduce a model for decision-making that comes from the US Air Force, called the **OODA loop**. This describes how a robot (or a person) makes decisions. Our robot will have two of these loops, an inner loop or **introspective loop**, and an outward-looking **environment sensor loop**. The lower, inner loop takes priority over the slower, outer loop, just as the autonomic parts of your body (such as the heartbeat, breathing, and eating) take precedence over your task functions (such as going to work, paying bills, and mowing the yard). This makes our system a type of **subsumption architecture**, a biologically inspired control paradigm named by Rodney Brooks of MIT, one of the founders of iRobot and Rethink Robotics, and the designer of a robot named Baxter.

The OODA Loop

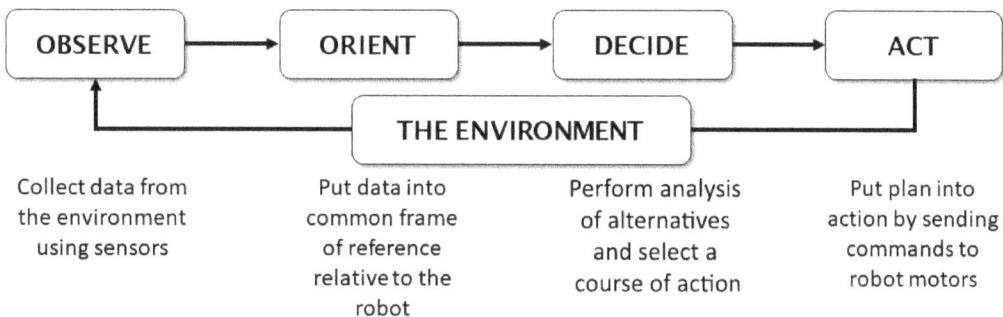

Figure 1.1 – My version of the OODA loop

> **Note**
>
> The OODA loop was invented by Col. John Boyd, a man also called *The Father of the F-16*. Col. Boyd's ideas are still widely quoted today, and his OODA loop is used to describe robot AI, military planning, or marketing strategies with equal utility. OODA provides a model for how a thinking machine that interacts with its environment might work.

Our robot works not by simply following commands or instructions step by step but by setting goals and then working to achieve those goals. The robot is free to set its own path or determine how to get to its goal. We will tell the robot *pick up that toy* and the robot will decide which toy, how to get in range, and how to pick up the toy. If we, the human robot owner, instead tried to treat the robot as a teleoperated hand, we would have to give the robot many individual instructions, such as *move forward*, *move right*, *extend arm*, *and open hand*, each individually, and without giving the robot any idea why we were making those motions. In a goal-oriented structure, the robot will be aware of which objects are toys and which are not and it will know how to find the toy box and how to put toys in the box. This is the difference between an autonomous robot and a radio-controlled teleoperated device.

Before designing the specifics of our robot and its software, we have to match its capabilities to the environment and the problem it must solve. The book will introduce some tools for designing the robot and managing the development of the software. We will use two tools from the discipline of systems engineering to accomplish this – **use cases** and **storyboards**. I will make this process as streamlined as possible. More advanced types of systems engineering are used by NASA, aerospace, and automobile companies to design rockets, cars, and aircraft – this gives you a taste of those types of structured processes.

The techniques used in this book

The following sections will each detail step-by-step examples of applying AI techniques to a robotics problem:

- We start with **object recognition**. We need our robot to recognize objects, and then classify them as either *toys* to be picked up or *not toys* to be left alone. We will use a trained **ANN** to recognize objects from a video camera from various angles and lighting conditions. We will be using the process of **transfer learning** to extend an existing object recognition system, **YOLOv8**, to recognize our toys quickly and reliably.

- The next task, once a toy is identified, is to pick it up. Writing a general-purpose *pick up anything* program for a robot arm is a difficult task involving a lot of higher mathematics (use the internet to look up *inverse kinematics* to see what I mean). What if we let the robot sort this out for itself? We use **genetic algorithms** that permit the robot to invent its own behaviors and learn to use its arm on its own. Then we will employ **deep reinforcement learning** (**DRL**) to let the robot teach itself how to grasp various objects using an **end effector** (robot speak for a hand).

- Our robot needs to understand commands and instructions from its owner (us). We use **natural language processing** (**NLP**) to not just recognize speech but to understand our intent for the robot to create goals consistent with what we want it to do. We use a neat technique that I call the *fill in the blank* method to allow the robot to reason from the context of a command. This process is useful for a lot of robot planning tasks.

- The robot's next problem is navigating rooms while avoiding the stairs and other hazards. We will use a combination of a unique, mapless navigation technique with 3D vision provided by a special stereo camera to see and avoid obstacles.

- The robot will need to be able to find the toy box to put items away, as well as have a general framework for planning moves in the future. We will use **decision trees** for path planning, as well as discussing **pruning** or quickly rejecting bad plans. If you imagine what a computer chess program algorithm must do, looking several moves ahead and scoring good moves versus bad moves before selecting a strategy, that will give you an idea of the power of this technique. This type of decision tree has many uses and can handle many dimensions of strategies. We'll be using it as one of two ways to find a path to our toy box to put toys away.

- Our final task brings a different set of tools not normally used in robotics, or at least not the way we are going to employ them.

  I have five wonderful, talented, and delightful grandchildren who love to come and visit. You'll be hearing a lot about them throughout the book. The oldest grandson is 10 years old, and autistic, as is my granddaughter, the third child, who is 8, as well as the youngest boy, who is 6 as I write this. I introduced my eldest grandson, William, to the robot – and he immediately wanted to have a conversation with it. He asked, *"What's your name?"* and *"What do you do?"* He was disappointed when the robot made no reply. So for the grandkids, we will be developing an engine for the robot to carry out a short conversation – we will be creating a robot personality to interact with children. William had one more request for this robot – he wants it to tell and respond to *knock, knock* jokes, so we will use that as a prototype of special dialog.

While developing a robot with actual feelings is far beyond the state of the art in robotics or AI today, we can simulate having a personality with a finite state machine and some Monte Carlo modeling. We will also give the robot a model for human interaction so that the robot will take into account the child's mood as well. I like to call this type of software an **AP** to distinguish it from our AI. AI builds a model of thinking, and an AP builds a model of emotion for our robot.

Now that you're familiar with the problem we will be addressing in this book, let's briefly discuss when and why you might need AI for your robot.

When do you need AI for your robot?

We generally describe AI as a technique for modeling or simulating processes that emulate how our brains make decisions. Let's discuss how AI can be used in robotics to provide capabilities that may be difficult for *traditional* programming techniques to achieve. One of those is identifying objects in images or pictures. If you connect a camera to a computer, the computer receives not an image, but an array of numbers that represent pixels (picture elements). If we are trying to determine whether a certain object, say a toy, is located in the image, then this can be quite tricky. You can find shapes, such as circles or squares, but a teddy bear? Moreover, what if the teddy bear is upside down, or lying flat on a surface? This is the sort of problem that an AI program can solve when nothing else can.

Our traditional approach for creating robot behaviors is to figure out what function we want and to write code to make that happen. When we have a simple function, such as driving around an obstacle, then this approach works well, and we can get results with a little tuning.

Some examples of AI and **ML** for robotics include:

- **NLP**: Using AI/ML to allow the robot to understand and respond to natural human speech and commands. This makes interacting with the robot much more intuitive.

- **Computer vision**: Using AI to let the robot see and recognize objects or people's faces, read text, and so on. This helps the robot operate in real-world environments.

- **Motion planning**: AI can help the robot plan optimal paths and motions to navigate around obstacles and people. This makes the robot's movements more efficient and human-like.

- **Reinforcement learning**: The robot can learn how to do, and improve at doing, tasks through trial and error using AI reinforcement learning algorithms. This means less explicit programming is needed.

The main rule of thumb is to use AI/ML whenever you want the robot to perform robustly in a complex, dynamic real-world environment. The AI gives it more perceptual and decision-making capabilities.

Now let's look at one function we need for this robot – recognizing that an object is either a toy (and needs to be picked up) or is not. Creating a standard function for this via programming is quite difficult. Regular computer vision processes separate an image into shapes, colors, or areas. Our problem is the toys don't have predictable shapes (circles, squares, or triangles), they don't have consistent colors, and they are not all the same size. What we would rather do is to teach the robot what is a toy and what is not. That is what we would do with a person. We just need a process for teaching the robot how to use a camera to recognize a particular object. Fortunately, this is an area of AI that has been deeply studied, and there are already techniques to accomplish this, which we will use in *Chapter 4*. We will use a **convolutional neural network** (**CNN**) to recognize toys from camera images. This is a type of **supervised learning**, where we use examples to show the software what type of object we want to recognize, and then create a customized function that *predicts* the class (or type) of object based on the pixels that represent it in an image. One of the principles of AI that we will be applying is **gradual learning** using **gradient descent**. This means that instead of trying to make the computer learn a skill all in one go, we will train it a little bit at a time, gently training a function to output what we want by looking at errors (or loss) and making small changes. We use the principle of gradient descent – looking at the slope of the change in errors – to determine which way to adjust the training.

You may be thinking at this point, *"If that works for learning to classify pictures, then maybe it can be used to classify other things,"* and you would be right. We'll use a similar approach – with somewhat different neural networks – to teach the robot to answer to its name, by recognizing the sound.

So, in general, when do we need to use AI in a robot? When we need to emulate some sort of decision-making process that would be difficult or impossible to create with procedural steps (i.e., programming). It's easy to see that neural networks are emulations of animal thought processes since they are a (greatly) simplified model of how neurons interact. Other AI techniques can be more difficult to understand.

One common theme could be that AI consistently uses *programming by example* as a technique to replace code with a common framework and variables with data. Instead of *programming by process*, we are programming by showing the software what result we want and having the software come up with how to get to that result. So for **object recognition** using pictures, we provide pictures of objects *and* the answer to what kind of object is represented by the picture. We repeat this over and over and train the software – by modifying the parameters in the code.

Another type of behavior we can create with AI has to do with behaviors. There are a lot of tasks that can be thought of as games. We can easily imagine how this works. Let's say you want your children to pick up the toys in their room. You could command them to do it – which may or may not work. Or, you could make it a game by awarding points for each toy picked up, and giving a reward (such as giving a dollar) based on the number of points scored. What did we add by doing this? We added a **metric**, or measurement tool, to let the children know how well they are doing – a point system. And, more critically, we added a reward for specific behaviors. This can be a process we can use to modify or create behaviors in a robot. This is formally called **reinforcement learning**. While we can't give a robot an emotional reward (as robots don't have wants or needs), we can program the robot to seek to maximize a reward function. Then we can use the same process of making a small adjustment in parameters that change the reward, see whether that improves the score, and then either keep that change (when learning results in more reward, our reinforcement) or discard it if the score goes down. This type of process works well for robot motion, and for controlling robot arms.

I must tell you that the task set out in this book – to pick up toys in an unstructured environment – is nearly impossible to perform without AI techniques. It could be done by modifying the environment – say, by putting RFID tags in the toys – but not otherwise. That, then, is the purpose of this book – to show how certain tasks, which are difficult or impossible to solve otherwise, can be completed using the combination of AI and robotics.

Next, let's discuss our robot and the development environment that we'll be using in this book.

Introducing the robot and our development environment

This is a book about robots and AI, so we really need to have a robot to use for all of our practical examples. As we will discuss in *Chapter 2* at some length, I have selected robot hardware and software that will be accessible to the average reader. The particular brand and type are not important, and I've upgraded Albert considerably since the first edition was published some five years ago. In the interest of keeping things up to date, we are putting all of the hardware details in the GitHub repository for this book.

As shown in the following photographs taken from two different perspectives, my robot has new omnidirectional wheels, a mechanical six-degree-of-freedom arm, and a computer brain:

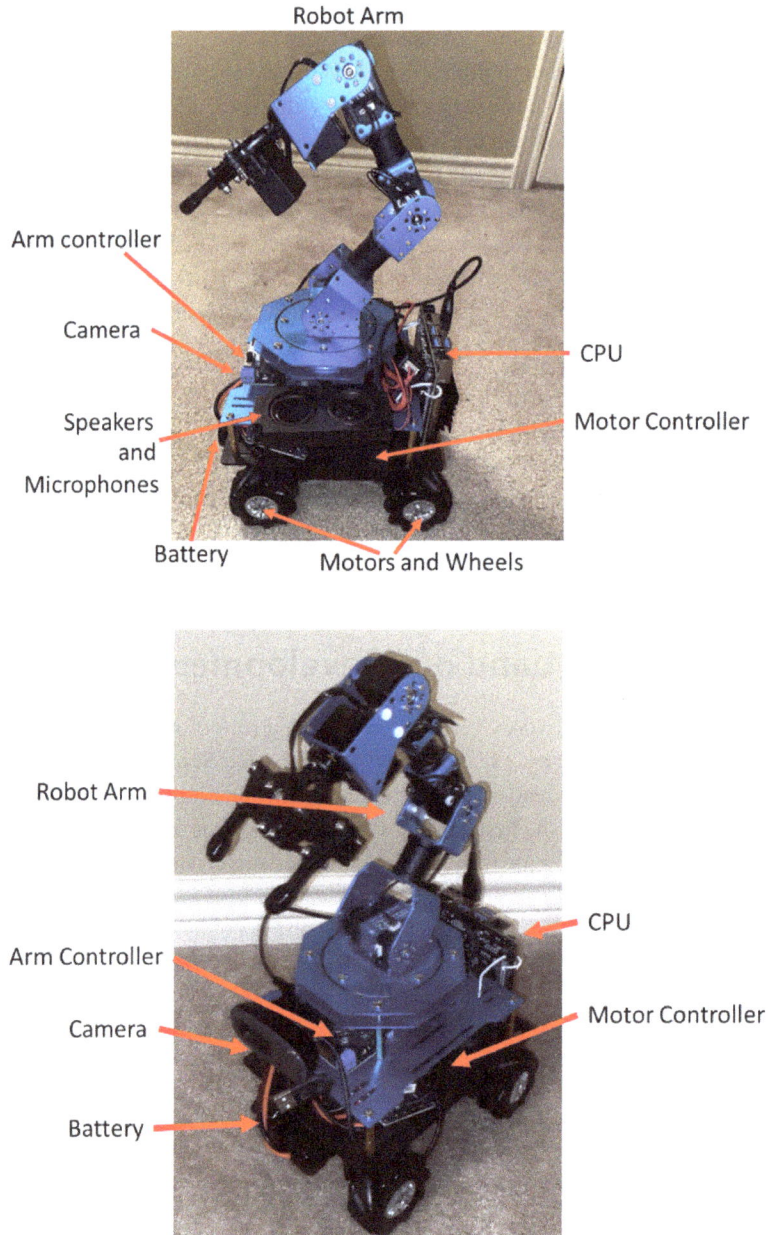

Figure 1.2 – Albert the robot has wheels and a mechanical arm

I'll call it *Albert*, since it needs some sort of name, and I like the reference to Prince Albert, consort of Queen Victoria, who was famous for taking marvelous care of their nine children. All nine of his children survived to adulthood, which was a rarity in the Victorian age, and he had 42 grandchildren. He went by his middle name; his actual first name was Francis.

Our tasks in this book center around picking up toys in an interior space, so our robot has a solid base with four motors and omni wheels for driving over carpet. Our steering method is the tank type, or differential drive, where we steer by sending different commands to the wheel motors. If we want to go straight ahead, we set all four motors to the same forward speed. If we want to travel backward, we reverse both motors the same amount. Turns are accomplished by moving one side forward and the other backward (which makes the robot turn in place) or by giving one side more forward drive than the other. We can make any sort of turn this way. The omni wheels allow us to do some other tricks as well – we can turn the wheels toward each other and translate directly sideways, and even turn in a circle while pointing at the same spot on the ground. We will mostly drive like a truck or car but will use the y-axis motion occasionally to line things up. Speaking of axes, I'll use the x axis to mean that the robot will move straight ahead, the y axis refers to horizontal movement from side to side, and the z axis is up and down, which we need for the robot's arm.

In order to pick up toys, we need some sort of manipulator, so I've included a six-axis robot arm that imitates a shoulder–elbow–wrist–hand combination that is quite dexterous and, since it is made out of standard digital servos, quite easy to wire and program.

The main control of the Albert robot is the Nvidia Nano **single-board computer** (**SBC**), which talks to the operator via a USB Wi-Fi dongle. The Nvidia talks to an Arduino Mega 2560 microcontroller and motor controller that we will use to control motors via **Pulse Width Modulation** (**PWM**) pulses. The following figure depicts the internal components of the robot:

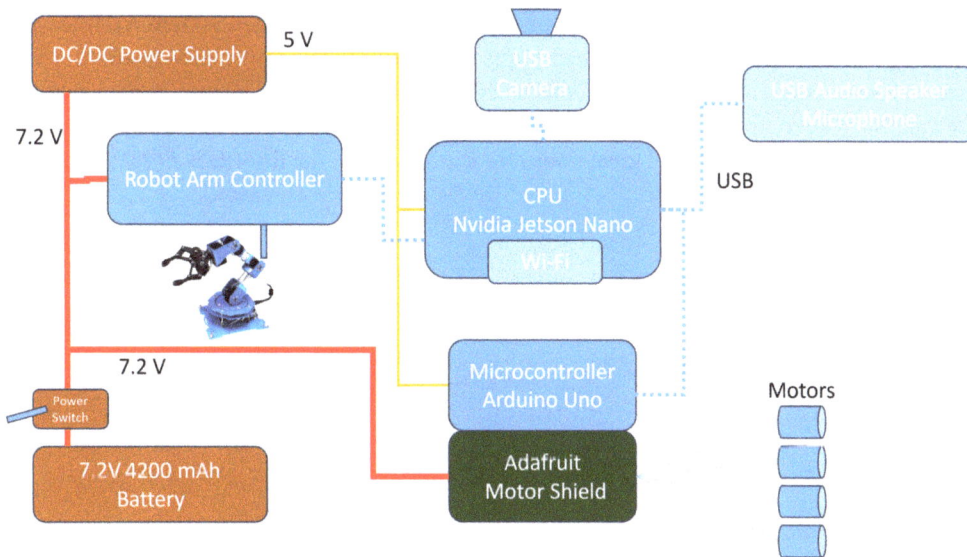

Figure 1.3 – Block diagram of the robot

We will be primarily concerned with the Nvidia Nano SBC, which is the brains of our robot. We will set up the rest of the components once and not change them for the entire book.

The Nvidia Nano acts as the main interface between our control station, which is a PC running Windows, and the robot itself via a Wi-Fi network. Just about any low-power, Linux-based SBC can perform this job, such as a BeagleBone Black, Odroid XU4, or an Intel Edison. One of the advantages of the Nano is that it can use its **Graphics Processing Units** (**GPUs**) to speed up the processing of neural networks.

Connected to the SBC is an Arduino with a motor controller. The Nano talks through a USB port addressed as a serial port. We also need a 5V regulator to provide the proper power from the 11.1V rechargeable lithium battery power pack into the robot. My power pack is a rechargeable 3S1P (three cells in series and one in parallel) 2700 Ah battery (normally used for quadcopter drones) and came with the appropriate charger. As with any lithium battery, follow all of the directions that come with the battery pack and recharge it in a metal box or container in case of fire.

Software components (ROS, Python, and Linux)

I am going to direct you once again to the Git repository to see all of the software that runs the robot, but I'll cover the basics here to remind you. The base operating system for the robot is Linux running on an Nvidia Nano SBC, as we said. We are using the ROS 2 to connect all of our various software components together, and it also does a wonderful job of taking care of all of the finicky networking tasks such as setting up sockets and establishing connections. It also comes with a great library of already prepared functions that we can just take advantage of, such as a joystick interface. ROS 2 is not a true operating system that controls the whole computer like Linux or Windows does, but rather is a backbone of communications and interface standards and utilities that make putting together a robot a lot simpler. The name I like to use for this type of system is **Modular Open System Architecture** (**MOSA**). ROS 2 uses a *publish/subscribe* technique to move data from one place to another that truly decouples the programs that produce data (such as sensors and cameras) from those programs that use data, such as controls and displays. We'll be making a lot of our own stuff and only using a few ROS functions. Packt has several great books for learning ROS; my favorite is *Effective Robotics Programming with ROS*.

The programming language we will use throughout this book, with a couple of minor exceptions, will be **Python**. Python is a great language for this purpose for two great reasons: it is widely used in the robotics community in conjunction with ROS, and it is also widely accepted in the machine learning and AI community. This double whammy makes using Python irresistible. Python is an interpreted language, which has three amazing advantages for us:

- **Portability**: Python is very portable between Windows, Mac, and Linux. Usually, you can get by with just a line or two of changes if you use a function out of the operating system, such as opening a file. Python has access to a huge collection of C/C++ libraries that also add to its utility.

- **No compilation**: As an interpreted language, Python does not require a compile step. Some of the programs we are developing in this book are pretty involved, and if we wrote them in C or C++, it would take 10 or 20 minutes of build time each time we made a change. You can do a lot with that much time, which you can spend getting your program to run and not waiting for the *make* process to finish.

- **Isolation**: This is a benefit that does not get talked about much but having had a lot of experience with crashing operating systems with robots, I can tell you that the fact that Python's interpreter is isolated from the core operating system means that having one of your Python ROS programs crash the computer is very rare. A computer crash means rebooting the computer and also probably losing all of the data you need to diagnose the crash. I had a professional robot project that we moved from Python to C++, and immediately the operating system crashes began, which shot the reliability of our robot. If a Python program crashes, another program can monitor that and restart it. If the operating system has crashed, there is not much you can do without some extra hardware that can push the *Reset* button for you.

Before we dive into the coding of our base control system, let's talk about the theory we will use to create a robust, modular, and flexible control system for robotics.

Robot control systems and a decision-making framework

As I mentioned earlier in this chapter, we are going to use two sets of tools in the sections: **soft real-time control** and the **OODA loop**. One gives us a base for controlling the robot easily and consistently, and the other provides a basis for the robot's autonomy.

How to control your robot

The basic concept of how a robot works, especially one that drives, is simple. There is a master control loop that does the same thing over and over – reads data from the sensors and motor controller, looks for commands from the operator (or the robot's autonomy functions), makes any changes to the state of the robot based on those commands, and then sends instructions to the motors or effectors to make the robot move.

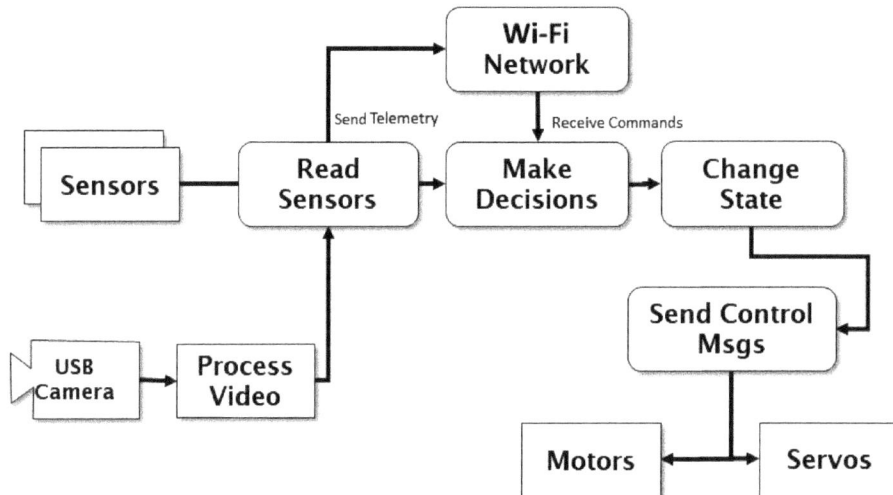

Figure 1.4 – Robot control loop

The preceding diagram illustrates how we have instantiated the OODA loop in the software and hardware of our robot. The robot can either act autonomously or accept commands from a control station connected via a wireless network.

What we need to do is perform this control loop in a consistent manner all of the time. We need to set a base frame rate or basic update frequency that sets the timing of our control loop. This makes all the systems of the robot perform together. Without some sort of time manager, each control cycle of the robot takes a different amount of time to complete, and any sort of path planning, position estimate, or arm movement becomes very complicated. ROS does not provide a time manager as it is inherently non-synchronous; if required, we have to create one ourselves.

Using control loops

In order to have control of our robot, we have to establish some sort of control or feedback loop. Let's say that we tell the robot to move 12 inches (30 cm) forward. The robot must send a command to the motors to start moving forward, and then have some sort of mechanism to measure 12 inches of travel. We can use several means, but let's just use a clock. The robot moves 3 inches (7.5 cm) per second. We need the control loop to start the movement, and then each update cycle, or time through the loop, check the time and see whether four seconds have elapsed. If they have, then it sends a *stop* command to the motors. The timer is the *control*, four seconds is the *set point*, and the motor is the *system* that is controlled. The process also generates an error signal that tells us what control to apply (in this case, to stop). Let's look at a simple control loop:

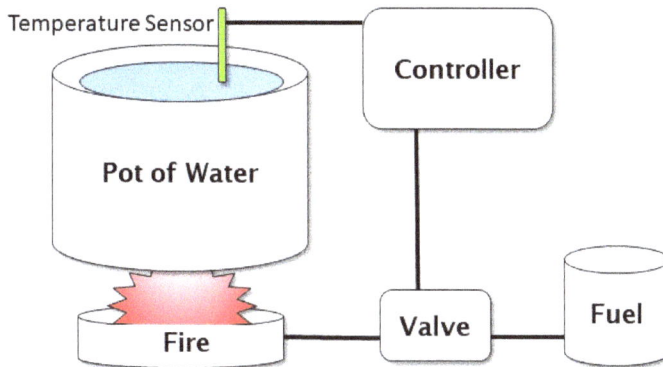

Figure 1.5 – Sample control loop – maintaining the temperature of a pot of water

Based on the preceding figure, what we want is a constant temperature in the **pot of water**. The **valve** controls the heat produced by the **fire**, which warms the **pot of water**. The **temperature sensor** detects whether the water is too cold, too hot, or just right. The controller uses this information to control the valve for more heat. This type of schema is called a **closed loop control system**.

You can think of this also in terms of a process. We start the process, and then get feedback to show our progress so that we know when to stop or modify the process. We could be doing speed control, where we need the robot to move at a specific speed, or pointing control, where the robot aims or turns in a specific direction.

Let's look at another example. We have a robot with a self-charging docking station, with a set of **light-emitting diodes** (**LEDs**) on the top as an optical target. We want the robot to drive straight into the docking station. We use the camera to see the target LEDs on the docking station. The camera generates an error signal, which is used to guide the robot toward the LEDs. The distance between the LEDs also gives us a rough range to the dock. This process is illustrated in the following figure:

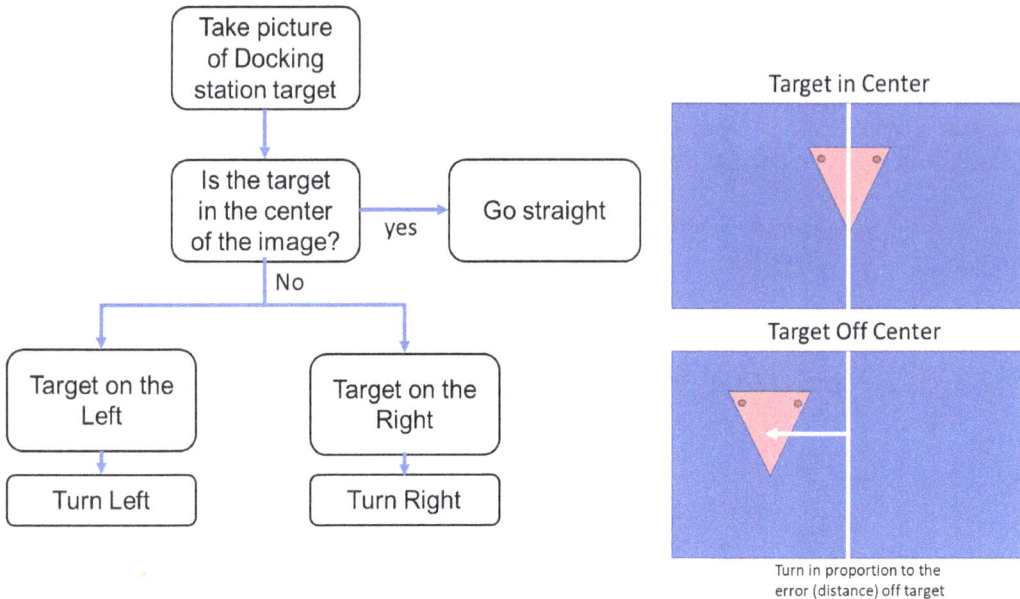

Figure 1.6 – Target tracking for a self-docking charging station

Let's understand this in some more detail:

- Let's say that the LEDs in the figure are off to the left of the center 50% and the distance from the robot to the target is 3 feet (1 m). We send that information through a control loop to the motors – turn left a bit and drive forward a bit.

- We then check again, and the LEDs are closer to the center (40%) and the distance to the target is 2.9 feet or 90 cm. Our error signal is a bit less, and the distance is a bit less. We'll have to develop a **scaling factor** to determine how many pixels equate to how much **turn rate**, which is measured as a percentage of full power. Since we are using a fixed camera and lens, this will be a constant.

- Now we send a slower turn and a slower movement to the motors this update cycle. We end up exactly in the center and come to zero speed just as we touch the docking station.

For those people currently saying, *"But if you use a PID controller …"*, yes, you are correct – you also know that I've just described a *P* or *proportional* control scheme. We can add more bells and whistles to help prevent the robot from overshooting or undershooting the target due to its own weight and inertia and to damp out oscillations caused by those overshoots.

A **PID controller** is a type of control system that uses three types of inputs to manage a closed-loop control system. A **proportional control** uses a multiple of the detected error to drive a control.

For example, in our pot of water, we measure the error in the temperature. If the desired temperature is 100°C and we measure 90°C with our thermometer, then the error in the temperature is 10 °C. We need to add more heat by opening the valve in proportion to the temperature error. If the error is 0, the change in the value is 0. Let's say that we try changing the value of the valve by 10% for a 10°C error. So we multiply 10°C by 0.01 to set our valve position to +0.1. This 0.01 value is our P term or **proportional constant**.

In our next sample, we see that the temperature of our pot is now 93°C and our error is 7°C. We change our valve position to +0.07, slightly less than before. We will probably find that by using this method, we will overshoot the desired temperature due to the hysteresis of the water – it takes a while for the water to heat up, creating a delay in the response. We will end up overheating the water and overshooting our desired temperature. One way to help prevent that is with the **D** term of the PID controller, that is, a **derivative** term. You remember that a derivative describes the slope of the line of a function – in this case, the temperature curve we measure. The y axis of our temperature graph is time, so we have *delta temperature/delta time*. To add a D term to our controller, we also add in the difference between the error of the last sample and the error of this sample *(-10 – (-7) = -3)*. We add this to our control by multiplying this value times a constant, D. The integral term is just the cumulative sum of the error multiplied by a constant we'll call I. We can modify the P, I, and D constants to adjust (tune) our PID controller to provide the proper response for our control loop – with no overshoots, undershoots, or drifts. More explanation is available at `https://jjrobots.com/pid/`. The point of these examples is to point out the concept of control in a machine – we have to take measurements, compare them to our desired result, compute the error signal, and then make any corrections to the controls over and over many times a second, and doing that consistently is the concept of real-time control.

Types of control loops

In order to perform our control loop at a consistent time interval (or to use the proper term, deterministically), we have two ways of controlling our program execution: **soft real time** and **hard real time**. Hard real-time control systems require assistance from the hardware of the computer – that is where the *hard* part of the title comes from. Hard real time generally requires a **real-time operating system (RTOS)** or complete control over all of the computer cycles in the processor. The problem we are faced with is that a computer running an operating system is constantly getting interrupted by other processes, chaining threads, switching contexts, and performing tasks. Your experience with desktop computers, or even smartphones, is that the same process, such as starting up a word processor program, always seems to take a different amount of time whenever you start it up.

This sort of behavior is intolerable in a real-time system where we need to know in advance exactly how long a process will take down to the microsecond. You can easily imagine the problems if we created an autopilot for an airliner that, instead of managing the aircraft's direction and altitude, was constantly getting interrupted by disk drive access or network calls that played havoc with the control loops giving you a smooth ride or making a touchdown on the runway.

An RTOS system allows the programmers and developers to have complete control over when and how the processes execute and which routines are allowed to interrupt and for how long. Control loops in RTOS systems always take the exact same number of computer cycles (and thus time) every loop, which makes them reliable and dependable when the output is critical. It is important to know that in a hard real-time system, the hardware enforces timing constraints and makes sure that the computer resources are available when they are needed.

We can actually do hard real time in an Arduino microcontroller because it has no operating system and can only do one task at a time or run only one program at a time. Our robot will also have a more capable processor in the form of an Nvidia Nano running Linux. This computer, which has some real power, does a number of tasks simultaneously to support the operating system, run the network interface, send graphics to the output HDMI port, provide a user interface, and even support multiple users.

Soft real time is a bit more of a relaxed approach, and is more appropriate to our playroom-cleaning robot than a safety-critical hard real-time system – plus, RTOSs can be expensive (there are open source versions) and require special training for you. What we are going to do is treat our control loop as a feedback system. We will leave some extra room – say about 10% – at the end of each cycle to allow the operating system to do its work, which should leave us with a consistent control loop that executes at a constant time interval. Just like our control loop example that we just discussed, we will take a measurement, determine the error, and apply a correction to each cycle.

We are not just worried about our update rate. We also must worry about **jitter**, or random variability in the timing loop caused by the operating system getting interrupted and doing other things. An interrupt will cause our timing loop to take longer, causing a random jump in our cycle time. We have to design our control loops to handle a certain amount of jitter for soft real time, but these are comparatively infrequent events.

Running a control loop

The process of running a control loop is fairly simple in practice. We start by initializing our timer, which needs to be the high-resolution clock. We are writing our control loop in Python, so we will use the `time.time()` function, which is specifically designed to measure our internal program timing performance (set frame rate, do loop, measure time, generate error, sleep for error, loop). Each time we call `time.time()`, we get a floating-point number, which is the number of seconds from the Unix clock and has microsecond resolution on the Nvidia Nano.

The concept for this process is to divide our processing into a set of fixed time intervals we will call **frames**. Everything we do will fit within an integral number of frames. Our basic running speed will process 30 **frames per second** (**fps**). That is how fast we will be updating the robot's position estimate, reading sensors, and sending commands to motors. We have other functions that run slower than the 30 frames, so we can divide them between frames in even multiples. Some functions run every frame (30 fps) and are called and executed every frame.

Let's say that we have a sonar sensor that can only update 10 times a second. We call the *read sonar* function every third frame. We assign all our functions to be some multiple of our basic 30 fps frame rate, so we have 30, 15, 10, 7.5, 6, 5, 4.28, 2, and 1 fps if we call the functions every frame, every second frame, every third frame, and so on. We can even do less than 1 fps – a function called every 60 frames executes once every 2 seconds.

The tricky bit is we need to make sure that each process fits into one frame time – which is 1/30 of a second or 0.033 seconds or 33 milliseconds. If the process takes longer than that, we have to either divide it up into parts or run it in a separate thread or program where we can start the process in one frame and get the result in another. It is also important to try and balance the frames so that not all processing lands in the same frame. The following figure shows a task scheduling system based on a 30 fps basic rate. Here, we have four tasks to take care of: task *A* runs at 15 fps, task *B* runs at 6 fps (every five frames), task *C* runs at 10 fps (every three frames), and task *D* runs at 30 fps (every frame):

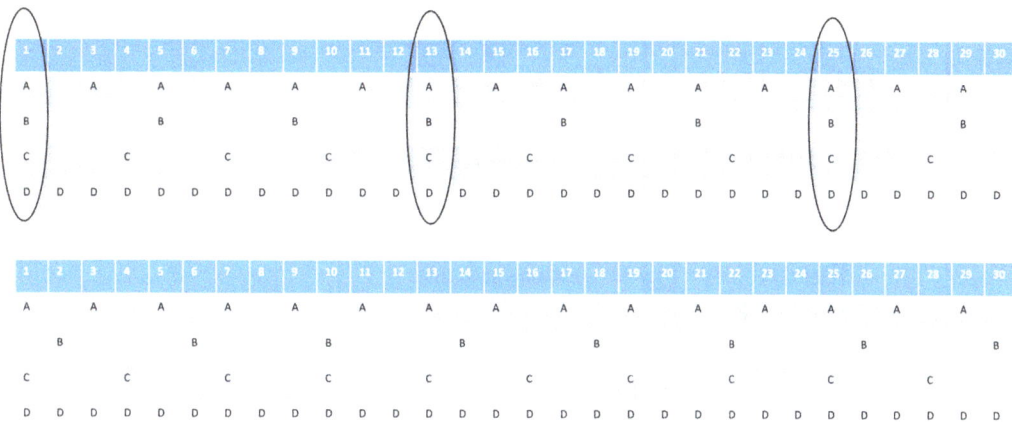

Figure 1.7 – Frame-based task schedule

Our first pass (the top of the figure) at the schedule has all four tasks landing on the same frame at frames 1, 13, and 25. We can improve the balance of the load on the control program if we delay the start of task *B* on the second frame as shown in the bottom half of the diagram.

This is akin to how measures in music work, where a measure is a certain amount of time, and different notes have different intervals – one whole note can only appear once per measure, a half note can appear twice, all the way down to 64th notes. Just like a composer makes sure that each measure has the right number of beats, we can make sure that our control loop has a balanced measure of processes to execute each frame.

Let's start by writing a little program to control our timing loop and to let you play with these principles.

This is exciting – our first bit of coding together. This program just demonstrates the timing control loop we are going to use in the main robot control program and is here to let you play around with some parameters and see the results. This is the simplest version I think is possible of a soft time-controlled loop, so feel free to improve and embellish it. I've made you a flowchart to help you understand this a little better:

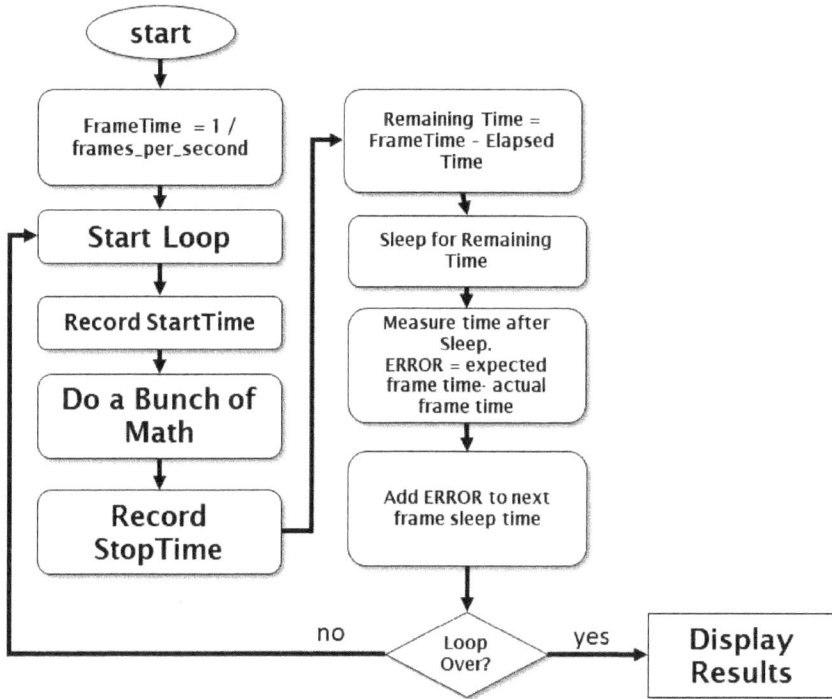

Figure 1.8 – Flowchart of soft real-time controller

Let's look more closely at the terms used in the preceding diagram:

- **FrameTime**: The time we have allocated to execute one iteration of the loop
- **StartTime**: When the loop/frame begins
- **Do a Bunch of Math**: The program that you are managing
- **StopTime**: When the frame completes
- **Remaining Time**: The difference between the elapsed time and the desired frame time
- **Elapsed Time**: The time it takes to actually run through the loop once
- **Frame Sleep Time**: We use **Remaining Time** to tell the computer to sleep so that the frame takes exactly the amount of time we want.

Now we'll begin with coding. This is pretty straightforward Python code – we won't get fancy until later:

1. We start by importing our libraries. It is not surprising that we start with the `time` module. We also will use the `mean` function from `numpy` (Python numerical analysis) and `matplotlib` to draw our graph at the end. We will also be doing some math calculations to simulate our processing and create a load on the frame rate:

```
import time
from numpy import mean
import matplotlib.pyplot as plt
import math
#
```

2. Now we have some parameters to control our test. This is where you can experiment with different timings. Our basic control is FRAMERATE – how many updates per second do we want to try? Let's start with 30, as we did in the example we discussed earlier:

```
# set our frame rate - how many cycles per second to run our
loop?
FRAMERATE = 30
# how long does each frame take in seconds?
FRAME = 1.0/FRAMERATE
# initialize myTimer
# This is one of our timer variables where we will store the
clock time from the operating system.
myTimer = 0.0
```

3. The duration of the test is set by the `counter` variable. The time the test will take is the FRAME time times the number of cycles in `counter`. In our example, 2,000 frames divided by 30 fps is 66.6 seconds, or a bit over a minute to run the test:

```
# how many cycles to test? counter*FRAME = runtime in seconds
counter = 2000
```

We will be controlling our timing loop in two ways:

- We will first measure the amount of time it takes to perform the calculations for this frame. We have a stub of a program with some trigonometry functions we will call to put a load on the computer. Robot control functions, such as computing the angles needed in a robot arm, need lots of trig to work. This is available from `import math` in the header of the program.

> **Note**
>
> We will measure the time for our control function to run, which will take some part of our frame. We then compute how much of our frame remains, and tell the computer to sleep this process for the rest of the time. Using the `sleep` function releases the computer to go and take care of other business in the operating system, and is a better way to mark time rather than running a tight loop of some sort to waste the rest of our frame time.

- The second way we control our loop is by measuring the complete frame – compute time plus rest time – and looking to see whether we are over or under our frame time. We use `TIME_CORRECTION` for this function to trim our sleep time to account for variability in the sleep function and any delays getting back from the operating system:

```
# factor for our timing loop computations
TIME_CORRECTION= 0.0
```

4. We will collect some data to draw a jitter graph at the end of the program. We use the `dataStore` structure for this. Let's put a header on the screen to tell you the program has begun, since it takes a while to finish:

```
# place to store data
dataStore = []
# Operator information ready to go
# We create a heading to show that the program is starting its
test
print "START COUNTING: FRAME TIME", FRAME, "RUN
TIME:",FRAME*counter
```

5. In this step, we are going to set up some variables to measure our timing. As we mentioned, the objective is to have a bunch of compute frames, each the same length. Each frame has two parts: a **compute** part, where we are doing work, and a **sleep** period, when we are allowing the computer to do other things. `myTime` is the *top of frame* time, when the frame begins. `newTime` is the end of the work period timer. We use `masterTime` to compute the total time the program is running:

```
# initialize the precision clock
 myTime = newTime = time.time()
 # save the starting time for later
 masterTime=myTime
 # begin our timing loop
 for ii in range(counter):
```

6. This section is our **payload** – the section of the code doing the work. This might be an arm angle calculation, a state estimate, or a command interpreter. We'll stick in some trig functions and some math to get the CPU to do some work for us. Normally, this *working* section is the majority of our frame, so let's repeat these math terms 1,000 times:

```
# we start our frame - this represents doing some detailed
math calculations
# this is just to burn up some CPU cycles
for jj in range(1000):
    x = 100
    y = 23 + ii
    z = math.cos(x)
    z1 = math.sin(y)
#
# read the clock after all compute is done
# this is our working frame time
#
```

7. Now we read the clock to find the working time. We can now compute how long we need to sleep the process before the next frame. The important part is that *working time + sleep time = frame time*. I'll call this `timeError`:

```
newTime = time.time()
# how much time has elapsed so far in this frame
# time = UNIX clock in seconds
# so we have to subract our starting time to get the elapsed
time
myTimer = newTime-myTime
# what is the time left to go in the frame?
timeError = FRAME-myTimer
```

We carry forward some information from the previous frame here. `TIME_CORRECTION` is our adjustment for any timing errors in the previous frame time. We initialized it earlier to zero before we started our loop so we don't get an undefined variable error here. We also do some range checking because we can get some large jitters in our timing caused by the operating system that can cause our sleep timer to crash if we try to sleep a negative amount of time:

> **Note**
>
> We use the Python `max` function as a quick way to clamp the value of sleep time to be zero or greater. It returns the greater of two arguments. The alternative is something like *if a< 0 : a=0.*

```
# OK time to sleep
# the TIME CORRECTION helps account for all of this clock
reading
# this also corrects for sleep timer errors
# we are using a porpotional control to get the system to
converge
# if you leave the divisor out, then the system oscillates
out of control
sleepTime = timeError + (TIME_CORRECTION/2.0)
# quick way to eliminate any negative numbers
# which are possible due to jitter
# and will cause the program to crash
sleepTime=max(sleepTime,0.0)
```

8. So, here is our actual sleep command. The `sleep` command does not always provide a precise time interval, so we will be checking for errors:

```
# put this process to sleep
time.sleep(sleepTime)
```

9. This is the time correction section. We figure out how long our frame time was in total (working and sleeping) and subtract it from what we want the frame time to be (`FrameTime`). Then we set our time correction to that value. I'm also going to save the measured frame time into a data store so we can graph how we did later using `matplotlib`. This technique is one of Python's more useful features:

```
#print timeError,TIME_CORRECTION
# set our timer up for the next frame
time2=time.time()
measuredFrameTime = time2-myTime
##print measuredFrameTime,
TIME_CORRECTION=FRAME-(measuredFrameTime)
dataStore.append(measuredFrameTime*1000)
#TIME_CORRECTION=max(-FRAME,TIME_CORRECTION)
#print TIME_CORRECTION
myTime = time.time()
```

This completes the looping section of the program. This example does 2,000 cycles of 30 frames a second and finishes in 66.6 seconds. You can experiment with different cycle times and frame rates.

10. Now that we have completed the program, we can make a little report and a graph. We print out the frame time and total runtime, compute the average frame time (total time/counter), and display the average error we encountered, which we can get by averaging the data in `dataStore`:

```
# Timing loop test is over - print the results
#
# get the total time for the program
endTime = time.time() - masterTime
# compute the average frame time by dividing total time by our
number of frames
avgTime = endTime / counter
#print report
 print "FINISHED COUNTING"
 print "REQUESTED FRAME TIME:",FRAME,"AVG FRAME TIME:",avgTime
 print "REQUESTED TOTAL TIME:",FRAME*counter,"ACTUAL TOTAL
TIME:", endTime
 print "AVERAGE ERROR",FRAME-avgTime, "TOTAL_
ERROR:",(FRAME*counter) - endTime
 print "AVERAGE SLEEP TIME: ",mean(dataStore),"AVERAGE RUN
TIME",(FRAME*1000)-mean(dataStore)
 # loop is over, plot result
 # this lets us see the "jitter" in the result
 plt.plot(dataStore)
 plt.show()
```

The results from our program are shown in the following code block. Note that the average error is just 0.00018 of a second, or 0.18 milliseconds out of a frame of 33 milliseconds:

```
START COUNTING: FRAME TIME 0.0333333333333 RUN TIME:
66.6666666667
FINISHED COUNTING
REQUESTED FRAME TIME: 0.0333333333333 AVG FRAME TIME:
0.0331549999714
REQUESTED TOTAL TIME: 66.6666666667 ACTUAL TOTAL TIME:
66.3099999428
AVERAGE ERROR 0.000178333361944 TOTAL_ERROR: 0.356666723887
AVERAGE SLEEP TIME: 33.1549999714 AVERAGE RUN TIME
0.178333361944
```

The following figure shows the timing graph of our program:

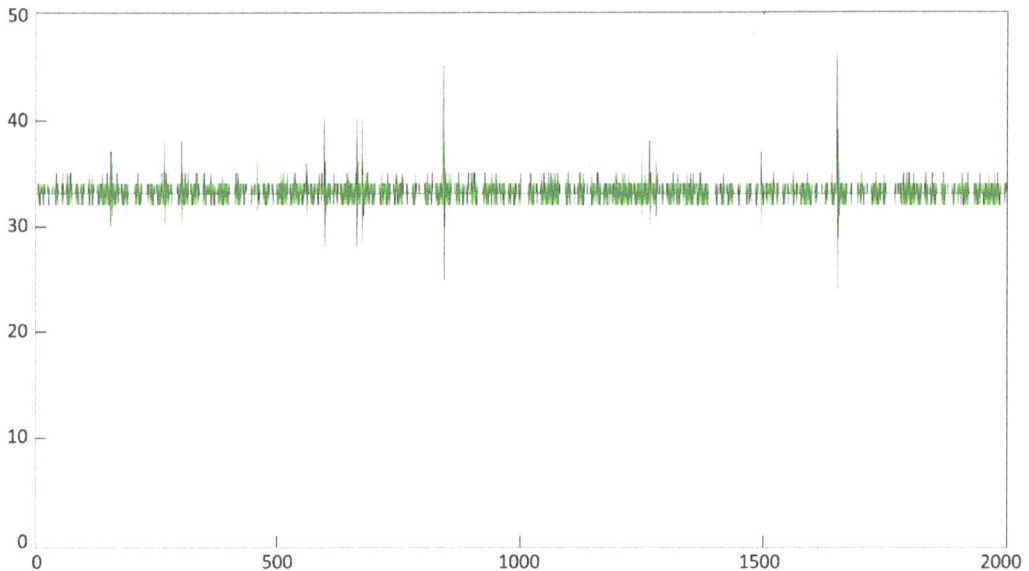

Figure 1.9 – Timing graph of our program

The *spikes* in the image are jitter caused by operating system interrupts. You can see the program controls the frame time in a fairly narrow range. If we did not provide control, the frame time would get greater and greater as the program executed. The graph shows that the frame time stays in a narrow range that keeps returning to the correct value.

Now that we have exercised our programming muscles, we can apply this knowledge to the main control loop for our robot with soft real-time control. This control loop has two primary functions:

- Respond to commands from the control station
- Interface to the robot's motors and sensors in the Arduino Mega

We will discuss this in detail in *Chapter 7*.

Summary

In this chapter, we introduced the subject of AI, which will be emphasized throughout this book. We identified the main difference between an AI robot and a *regular* robot, which is that an AI robot may be nondeterministic. This is to say it may have a different response to the same stimulus, due to learning. We introduced the problem we will use throughout the book, which is picking up toys in a playroom and putting them into a toy box. Next, we discussed two critical tools for AI robotics: the OODA loop, which provides a model for how our robot makes decisions, and the soft real-time control loop, which manages and controls the speed of execution of our program. We applied these techniques in a timing loop demonstration and began to develop our main robot control program.

In the next chapter, we will teach the robot to recognize toys – the objects we want the robot to pick up and put away. We will use computer vision with a video camera to find and recognize the toys left on the floor.

Questions

1. What does the acronym *PID* stand for? Is this considered an AI software method?

2. What is the Turing test? Do you feel this is a valid method of assessing AI?

3. Why do you think robots have a problem with negative obstacles such as stairs and potholes?

4. In the OODA loop, what does the *Orient* step do?

5. From the discussion of Python and its advantages, compute the following. Your program needs 50 changes tested. Assuming each change requires a recompile step and one run to test, a C Make compile takes 450 seconds and a Python `run` command takes 3 seconds. How much time do you sit idle waiting on the compiler?

6. What does RTOS stand for?

7. Your robot has the following scheduled tasks: telemetry at 10 Hz, GPS at 5 Hz, inertial measurements at 50 Hz, and motor control at 20 Hz. At what frequency would you schedule the base task, and what intervals would you use for the slower tasks (i.e., 10 Hz base, motors every three frames, telemetry every two frames, etc.)?

8. Given that a frame rate scheduler has the fastest task at 20 fps, how would you schedule a task that needs to run at 7 fps? How about one that runs at 3.5 fps?

9. What is a blocking call function? Why is it bad to use blocking calls in a real-time system like a robot?

Further reading

You can refer to the following resources for further details:

- *Effective Robotics Programming with ROS – Third Edition*, by Anil Mahtani, Luis Sanchez, and Enreque Fernandez Perdomo, Packt Publishing, 2016

- *Introduction to AI Robotics – Second Edition* by Robin R. Murphy, Bradford Books, 2019

- *Real-Time scheduling: from hard to soft real-time systems*, a whitepaper by Palopoli Lipari, 2015 (https://arxiv.org/pdf/1512.01978.pdf)

- *Boyd: The Fighter Pilot Who Changed the Art of War*, by Robert Coram, Little, Brown and Company, 2002

2
Setting Up Your Robot

This chapter begins with some background on my thoughts on what a robot is, and what robots are made of – a fairly standard list of parts and components. This chapter aims to allow you to duplicate the exercises and use the source code that is found throughout the book. I will describe how I set up my environments for development, what tools I used to create my code, and how to install the **Robotic Operating System version 2** (**ROS 2**). The assembly of Albert, the robot I use for all the examples, is covered in the GitHub repository for this book. There are many other types and configurations of robots that can work with the code in this book with some changes. I'll try to provide all the shortcuts I can, including a full image of my robot's SD card, in the Git repo.

In this chapter, we will be covering the following topics:

- Understanding the anatomy of a robot

- Introducing subsumption architecture

- A brief introduction to ROS

- Software setup: Linux, ROS 2, Jetson Nano, and Arduino

Technical requirements

To complete the practical exercises in this chapter, you will need the requirements specified in the *Preface* at the beginning of this book. The code for this chapter can be found at https://github.com/PacktPublishing/Artificial-Intelligence-for-Robotics-2e/.

Understanding the anatomy of a robot

A robot is a machine that is capable of carrying out complex actions and behaviors by itself. Most robots are controlled by a computer or digital programmable device. Some key characteristics of robots are as follows:

- **Automation**: Robots can operate automatically without direct human input, based on their programming. This allows them to do repetitive or dangerous tasks consistently.

- **Sensors**: Robots use sensors such as cameras, optics, lidar, and pressure sensors to gather information about their environment so they can navigate and interact. This sensory information is processed to determine what actions the robot should take.

- **Programming**: A robot's *brain* consists of an onboard computer or device that runs code and algorithms that define how it will behave. Robots are programmed by humans to perform desired behaviors.

- **Movement**: Most robots are able to move around to some degree through wheels, legs, propellers, or other locomotion systems. This allows them to travel through environments to perform tasks.

- **Interaction**: Advanced robots can communicate with humans through voice, visual displays, lights, sounds, physical gestures, and more. This allows useful human-robot interaction and work.

- **Autonomy**: While robots are programmed by humans, they have a degree of self-governance and independence in how they meet their objectives. The ability to take action and make decisions without human oversight is their autonomy.

In summary, a robot integrates automation, sensing, movement, programming, and autonomy to reliably carry out jobs that may be complex, repetitive, unsafe, or otherwise unsuitable for humans. They come in many shapes and sizes, from industrial robotic arms to social companion robots to autonomous self-driving cars.

There is a fairly standard collection of components and parts that make up the vast majority of robots. Even robots as outwardly different as a self-driving car, the welding robot that built the car, and a Roomba vacuum cleaner have a lot of the same components or parts. Some will have more, and some will have less, but most mobile robots will have the following categories of parts:

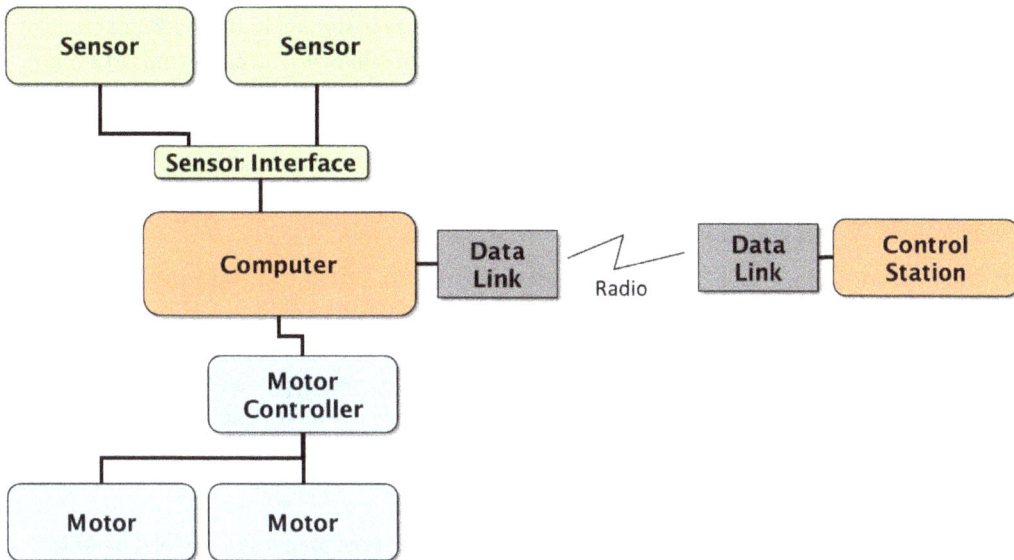

Figure 2.1 – Block diagram of a typical mobile robot

Let's look at these components in greater detail:

- **Computer**: A unit that runs the programming that controls the robot. This can be a traditional computer, a microcontroller, a **single board computer** (**SBC**) like we have, or some other sort of processor that sends and receives commands. Robot arms and some types of industrial robots will use a **Programmable Logic Controller** (**PLC**), which is a special type of controller that applies logic (*AND, OR, NOT*) to various inputs to produce an output. For a computer to send commands to the robot and receive telemetry, we'll need some sort of sensor interface, such as a USB port, serial port, **General Purpose Input/Output** (**GPIO**) port, or a network interface such as Ethernet or Wi-Fi.

- **Control Station** or **Human/Robot Interface** (**HRI**): Robots are designed to perform tasks, which requires that the operator must have some means to send and receive data from the robot and to supervise that the robot is behaving correctly. We will be using a laptop or desktop computer for this function, and we will talk to the robot via a wireless network. Our control station sends commands to the robot and receives **telemetry** from the robot in the form of **data**, **video**, or **audio**.

- **Radio** or **Data Link**: Mobile robots such as the one we are designing in this book are capable of moving and exploring their environment. While it is possible to send commands to a robot over a tether or wire, the preferred way is to use a radio link. The ubiquitous availability of wireless networks such as Wi-Fi and cellular data services has made creating data links a lot easier. I have had a lot of robot projects where a network link was unavailable or impractical, and a custom radio solution needed to be devised. Other types of radio used in robots include Bluetooth, Zigbee, and various mesh network systems, such as Flutter.

- **Motors** or **Effectors**: Our definition of a robot includes the ability for **self-propulsion**; that is, the robot is able to move. In order to move, the robot needs a motor or set of motors. Our robot, Albert, has ten motors, four for driving and six to control the robot arm and hand. Motors convert electricity into motion. There are many different types, and picking the right motor is a challenge. You must match the torque (how hard the motor can pull), the speed of the motor shaft in revolutions per minute, and voltage. Here are some key factors to consider when selecting a motor for a robot drive system:

  - **Torque**: Consider the torque required for your robot's movements and payload handling. More torque allows faster acceleration and the ability to handle heavier loads. If there is insufficient torque, the robot will "bog down" or stall the motor. An electric motor pulls the most current when it is stalled (it is energized but not moving). All that power going nowhere gets turned into heat, which will eventually melt the wires or cause a fire.

  - **Speed**: Determine the speeds your robot needs to operate at. Higher speeds require motors with higher RPM ratings. We only want our robot to go at a modest rate. The toys can't get away.

  - **Duty cycle**: Choose a motor that can run continuously for the robot's required duty cycle without overheating. Intermittent duty cycles allow smaller, lighter motors. We will be driving or moving quite a bit – about 50% of the time, but not too fast.

  - **Size and weight**: Large, heavy-duty motors provide a lot of power but may constrain robot design. Consider the full drive system size and weight. Remember the motor also has to move itself.

  - **Control**: Brushless DC motors require electronic speed controllers. Stepper motors allow open-loop position control. Servomotors, such as the ones in the robot's arm, have integrated encoders and are controlled by a serial interface. The drive motors I used are brushed motors, which are controlled by varying the voltage, which we control with **Pulse Width Modulation** (**PWM**).

  - **Voltage**: High voltages allow more power in small motors. Select a voltage that is compatible with other electronics. My battery is 7.2 volts, which matches the motors selected.

  - **Noise**: Quiet motors may be required for home/office robots. Brushless, gear-reduced motors are quiet but expensive. Geared drivetrains are also noisy.

  - **Cost**: More powerful motors cost more. Balance performance needs with budget constraints. Albert's brushed motors are very inexpensive.

Some robot motors also feature gearboxes to reduce the motor speed, basically exchanging speed for torque. Albert's electric motors have reduction gearboxes that let the motor run at a faster speed than the wheels.

There are many ways to provide motion to a robot. We call these *things that make the robot move* **effectors**. Effectors are only limited by your imagination, and include **pneumatics** (things actuated by compressed air), **hydraulics** (things actuated by incompressible fluid), **linear actuators** (things that convert rotary motion into linear motion), **revolving joints** or **revolute joints** (angular joints like an elbow) and even exotic effectors such as **shape-memory alloy** or **piezoelectric crystals**, which change shape when electricity is applied.

- **Servos**: Some of the motors in our robot are a special category of motors called **servos**. Servo motors feature a feedback mechanism and a control loop, either to maintain a position or a speed. The feedback is provided by some sort of **sensor**. The servos we are using consist of a small electric motor that drives a gearbox made up of a series of gears that reduce the speed and consequently increase the torque of the motor. The sensor used in our case is a potentiometer (variable resistor) that can measure the angle of the output gear shaft. When we send a command to the servo, it tells the motor to be set to a particular angle. The angle is measured by the sensor, and any difference between the motor's position and the sensor creates an error signal that moves the motor in the correct direction. You can hear the motor making a lot of noise because the motor turns many times through seven reduction gears to make the arm move. The gearbox lets us get a lot of torque without drawing a lot of current.

Figure 2.2 shows how a servo motor is controlled using **Pulse Position Modulation** (PPM). To control a servo, you must generate a pulse of a specific width:

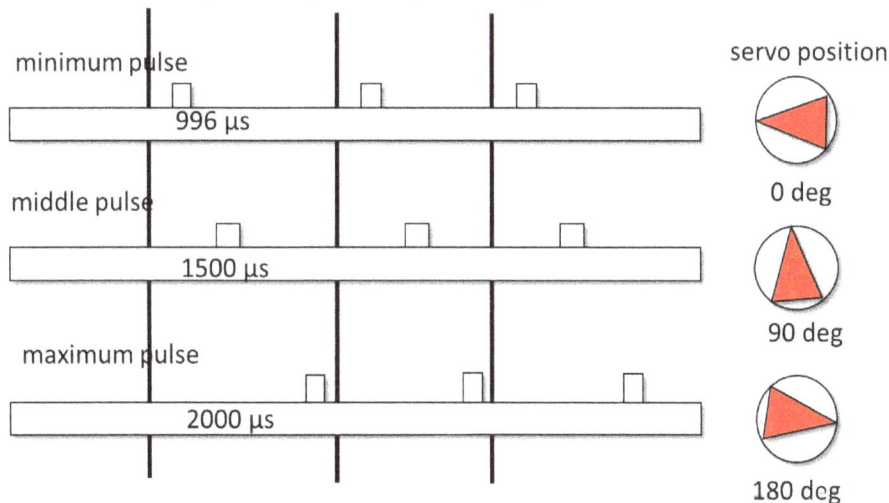

Figure 2.2 – Servo motor control is via PPM signals

A short pulse moves the servo to the beginning of its range. A medium pulse (1,500 microseconds) is the center of the servo's position. A late pulse causes the servo motor to go to the end of its range. The robot arm I use in this version of my robot has a servo controller that comes with the arm hardware. We will be controlling the robot arm via serial commands to this controller in *Chapter 5*.

- **Motor Controller or Electronic Speed Control**: Motors are not very useful by themselves – you need the ability to convert commands from the control computer into motion from the motors. Since motors need more voltage and more current than the control computer (our Jetson Nano) can provide, we need a device to turn small digital signals into large analog voltage and current. This device is called a motor controller. This controller I had to purchase separately, and is composed of two parts – an Arduino Uno and a motor controller shield that is attached:

Figure 2.3 – Motor controller shield I used for Albert

As shown in the image, the four motor wires are attached to the lettered connections.

- Since we have a tank-drive robot (we steer by running the motors at different speeds, also called **differential drive**), we also need the motors to be able to run forward or backward. The motor controller takes a special input signal called a **Pulse Width Modulation** (**PWM**). PWM is a repeating signal where the voltage turns on and off. The motor throttle (how fast the motor turns) is proportional to the amount the PWM signal stays in the *ON* position.

  The motor controller has several kinds of connections, and has to be wired carefully due to the high voltages and currents provided. This can be done by performing the following steps:

  - There are two **control wire inputs** – one for speed (the PWM signal) and the other is a direction signal. We put the motor in reverse by changing the **direction signal** – 1 is forward, and 0 is backward.

  - The next thing we need is a **ground** – it is very important that the controller sending the PWM signal (in our case, it is the Ardunio Mega) and the motor control have their ground lines connected.

  - Next, the motor controller needs the **motor voltage** and **current**, which we get directly from our battery.

  - Finally, we connect two wires from each motor to the controller. It is interesting that we don't care which wire goes to which side of the motor, since we can run both forward and backward. If the motor is turning the wrong way, just switch the two wires. This is the only time you get to say *just reverse the polarity* outside of a science fiction movie.

  We will be covering the specific wiring for the example robot – Albert – in the online appendix.

- **Sensors**: In order for the robot, which is a machine that can move and react to its environment, to be able to see its surroundings, it needs sensors. Sensors take information from the outside or inside of the robot and convert it into a digital form. If we use a digital **camera sensor**, it takes light and turns it into digital pixels (picture elements), recorded as an array of numbers. A **sonar sensor** measures the distance to an object, such as a wall, by sending a pulse of energy (sound waves) and listening for the time delay before hearing an echo. Measuring the time delay gives us the distance to an object, since the speed of sound is fairly constant. For our Albert project, the robot has several types of sensors:

  - Our **primary sensor** is a wide-angle video camera, which we will use for avoiding obstacles and detecting objects.

  - We will also use a **microphone** to listen for sounds and perform speech recognition.

  - We mentioned servo motors earlier in this list – each servo motor contains an **angle sensor** that detects the amount of rotation and allows us to direct the robot arm and hand.

- We have our **Emergency Stop button**, which is wired to the Arduino, and is a type of tactile (touch) sensor. When the button is pressed, a signal is sent that the robot can interpret as a stop command.

- The robot arm I chose has a handy **voltage monitor** that we will use to keep track of the battery life (charge) remaining.

In the next section, we will discuss robot software architectures, which act as a framework for the autonomy behaviors we will be creating.

Introducing subsumption architecture

At this point, I want to spend a bit of time on the idea behind the **subsumption architecture**, and point out some specifics of how we will be using this concept in the design of our robot project. Many of you will be familiar with the concept from school or from study, so you can look at my diagram and then move on. For the rest of us, let's talk a bit about this biologically inspired robot concept.

Subsumption architecture was originally described by Dr. Rodney Brooks, a professor at MIT, who would later help found iRobot Corporation and invent the Baxter robot. Rodney was trying to develop analogs of insect brains in order to understand how to program intelligent robots. Robots before this time (1986) were very much single-threaded machines that pretty much only did one thing at a time. They read sensors, made decisions, and then acted – and only had one goal at any one time. Creatures such as flies or ants have very simple brains but still manage to function in the real world. Brooks reasoned that there were several layers of closed-loop feedback processes going simultaneously.

The basic concept of subsumption has been around for some time, and it has been adapted, reused, refined, and simplified in the years since it was first introduced. What I am presenting here is my interpretation of how to apply the concept of subsumption to a robot in the context of what we are trying to accomplish.

The first aspect to understand is that we want our robot to act on a series of goals. The robot is not simply reacting to each stimulus in total isolation, but is rather carrying out some sort of goal-oriented behavior. The goal may be to pick up a toy or navigate the room, avoiding obstacles. The paradigm we are creating has the user set goals for the robot and the robot determines how to carry those goals out, even if the goal is simply to move one meter forward.

The problem begins when the robot has to keep more than one goal in mind at a time. The robot is not just driving around, but driving around avoiding obstacles and looking for toys to pick up. How do we arbitrate between different goals, to determine which one has precedence? The answer is found in the following diagram:

Subsumption Architecture

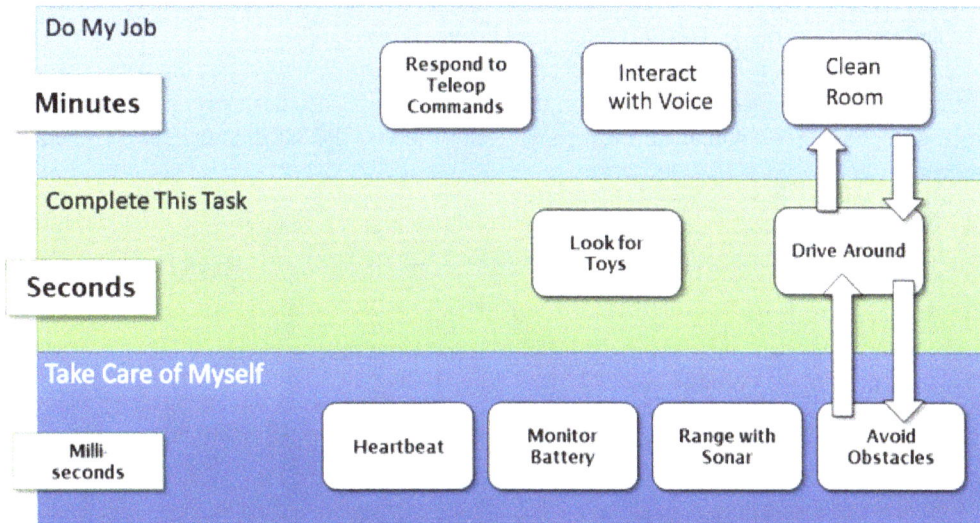

Figure 2.4 – An example subsumption architecture

We will divide the robot's decision-making systems into three layers, each of which has a different level of responsibility and operates on a different time scale.

At the lowest levels are what we might call the robot's autonomic nervous system – it contains the robot's internal health-keeping and monitoring functions. These processes run very fast – 20 times a second or so, or 20 hertz (Hz), and only deal with what is inside the robot. This includes reading internal sensors, checking battery levels, and reading and responding to heartbeat messages. I've labeled this level *Take Care of Myself*.

> **Important note**
>
> What is a **heartbeat message**? Once a second, I have the control station send a special heartbeat message to the robot, which has a time tag down to the millisecond, which is the clock time of the host computer. This goes to the control computer and repeats the heartbeat message back to the host. We can see the delay in our message – our command latency – by comparing the time tags. We want to see a less than 25 milliseconds round trip for the heartbeat. If the onboard computer is not working or is locked up, then the time tag won't come back and we know the robot is having problems.

The next level handles individual tasks, such as driving around or looking for toys. These tasks are short-term and deal with what the sensors can see. The time period for decisions is in the second range, so these tasks might have 1 or 2 Hz update rates, but slower than the internal checks. I call this level *Complete the Task* – you might call it *Drive the Vehicle* or *Operate the Payload*.

The final and top level is the section devoted to *completing the mission*, and it deals with the overall purpose of the robot. This level has the overall state machine for finding toys, picking them up, and then putting them away, which is the mission of this robot. This level also deals with interacting with humans and responding to commands. The top level works on tasks that take minutes, or even hours, to complete.

The rules of the subsumption architecture – and even where it gets its name – have to do with the priority and interaction of the processes in these layers. The rules are as follows (and this is my version):

- Each layer can only talk to the layers next to it. The top layer talks only to the middle layer, and the bottom layer also talks only to the middle layer. The middle layer can communicate with both the top and the bottom layer.

- The layer with the lower level has the highest priority. The lower level has the ability to interrupt or override the commands from higher layers.

Think about this for a minute. I've given you an example of driving our robot in a room. The lowest level detects obstacles. The middle level is driving the robot in a particular direction, and the top layer is directing the mission. From the top down, the uppermost layer is commanded to *clean up the room*, the middle layer is commanded to *drive around*, and the bottom layer gets the command *left motor and right motor forward 60% throttle*. Now, the bottom level detects an obstacle. It interrupts the *drive around* function and overrides the command from the top layer to turn the robot away from the obstacle. Once the obstacle is cleared, the lowest layer returns control to the middle layer for the driving direction.

Another example could be if the lowest layer loses the heartbeat signal, which indicates that something has gone wrong with the software or hardware. The lowest layer causes the motors to halt, overriding any commands from the upper layers. It does not matter what they want; the robot has a fault and needs to stop. This **priority inversion** of the lowest layers having the highest priority is the reason we call this a subsumption architecture, since the higher layers subsume – incorporate – the functions of the lower layers to perform their tasks.

The major benefit of this type of organization is that it keeps procedures clear as to which events, faults, or commands take precedence over others, and prevents the robot from getting stuck in an indecision loop.

Each type of robot may have different numbers of layers in their architecture. You could even have a **supervisory layer** that controls a number of other robots and has goals for the robots as a team. The most I have had so far has been five, used in one of my self-driving car projects.

Now let's take a look at one of the most important concepts you'll need in this book – ROS.

A brief introduction to ROS

OK, before we do all of the work described in the following section to be able to use ROS 2 – the second version of the Robotic Operating System – let's answer your questions. What is ROS, and what are its advantages?

The first thing to know is that ROS is not an actual operating system, such as Linux or Windows. Rather it is a middleware layer that serves as a means of connecting different programs to work together to control a robot. It was originally designed to run Willow Garage's PR2 robot, which was complex indeed. ROS is supported by a very large open source community and is constantly updated.

I used to be a ROS skeptic, and frankly, reading the documentation did not help my first impression that it was cumbersome at best and difficult to use. However, at the insistence of one of my business partners, we started using ROS for a very complex self-guided security guard robot called RAMSEE, designed for Gamma 2 Robotics:

Figure 2.5 – RAMSEE, the security guard robot, designed by the author

I quickly realized that while the initial learning curve with ROS was steep, the payoff was the ability to create and implement modular, easily portable services that could be developed independently. I did not need to combine everything into one program, or even in one CPU. I could take advantage of my multi-core computers to run independent processes, or even have more than one computer and move things freely from one to the other. RAMSEE has one computer with eight cores and another with four.

> **Important note**
>
> ROS can be described as a **Modular Open System Software** (MOSA). It provides a standard interface to allow programs to talk to one another through a *Publish-Subscribe* paradigm. This means that one program publishes data, making it available to other programs. The programs that need this subscribe to that data and are sent a message whenever new data is available. This lets us develop programs independently and create standardized interfaces between programs. It really makes creating robots much easier and far more flexible.

The other major advantage, and worth all the bother, is that ROS has a very large library of ready-to-go interfaces for sensors, motors, drivers, and effectors, as well as every imaginable type of robot navigation and control tool. For example, we will be using the OAK-D 3D depth camera, which has a ROS 2 driver available at `https://github.com/luxonis/depthai-ros`.

The RViz2 tool provides visualization of all of your sensor data, as well as showing the localization and navigation process. I greatly appreciated the logging and debugging tools included in ROS. You can log data – anything that crosses the publish/subscribe interface – to a **ROSBag** and play it back later to test your code without the robot being attached, which is very useful.

The following illustration below shows the output given by RViz2, showing a map being drawn by one of my robots:

Figure 2.6 – ROS RViz allows you to see what the robot sees, in this case, a map of a warehouse

Since this is the second edition of the book, we will be using ROS 2, the new and improved version of ROS. One of the most frustrating things about the old ROS was the use of **ROSCORE**, a traffic cop that connected all of the parts of the robot via the network. That is now gone, and the various components can find each other via a different sort of service, called **Distributed Data Services** (**DDS**). We will also need to use Python 3 instead of Python 2 for our code since Python 2 has been discontinued and is no longer supported.

Hardware and software setup

To match the examples in this book, and to have access to the same tools that are used in the code samples, you will have to set up three environments:

- **A laptop or desktop computer**: This will run our control panel, and also be used to train neural networks. I used a Windows 10 computer with Oracle VirtualBox supporting a virtual machine running Ubuntu 20.04. You may run a computer running Ubuntu or another Linux operating system by itself (without Windows) if you want. Several of the AI packages we will use in the tutorial sections of the book will require Ubuntu to run. We will load ROS 2 on this computer. I will also be using a PlayStation game controller on this computer for teleoperation (remote control) of the robot when we teach the robot how to navigate. I also have ROS 2 for Windows installed, which may obviate running the virtual machine. Either approach will work, since the Python programs we will use for control run in either mode.

- **Nvidia Jetson Nano 8GB**: This also runs Ubuntu Linux 20.04 (you can also run other Linux versions, but you will have to make any adjustments between those versions yourself). The Nano also runs ROS 2. We will cover the additional libraries we need in the following sub-sections.

- **Arduino Mega 256**: We need to be able to create code for the Arduino. I'm using the regular Arduino IDE from the Arduino website. It can be run on Windows or Linux. We will be using the Arduino to control the motors on the robot base and drive it around. It also gives us a lot of expansion to add additional controls, such as an emergency stop button.

Preparing the laptop

You will need to install ROS 2 for Windows for the robot control software to work. To do this, you can follow the instructions provided at `https://docs.ros.org/en/foxy/Installation/Windows-Install-Binary.html`.

I also used **Virtual Network Computing** (**VNC**) to talk to my Nano from the laptop, which saves a lot of time and fiddling with cables and keyboards. Otherwise, you would need to connect the Nano to a monitor, keyboard, and mouse to be able to work on your code that is on the robot. I used **RealVNC**, which can be found at `https://www.realvnc.com/en/`. You can also use **UltraVNC**, which is free software.

Installing Python

The Linux Ubuntu system will come with a default version of Python. I am going to assume that you are familiar with Python, as we will be using it throughout the book. If you need help with Python, Packt has several fine books on the subject.

Once you log on to your virtual machine, check which version of Python you have by opening a terminal window and typing `python` in the command prompt. You should see the Python version, like this:

```
>python
Python 3.8.16 (default, Jan 17 2023, 22:25:28) [MSC v.1916 64 bit
(AMD64)]
```

You can see that I have version 3.8.16 in this case.

We are going to need several add-on libraries that add on to Python and extend its capabilities. The first thing to check is to see if you have `pip` installed. This is the **Python Installation Package** (**PIP**) that loads other packages from the internet to extend Python. Check to see if you have `pip` by typing in the following:

```
pip
```

If you get the output `No command 'pip' found`, then you need to install Pip. Enter the following:

```
sudo apt-get install python-pip python-dev build-essential
sudo pip install --upgrade pip
```

Now we can install the rest of the packages we need. As a start, we need the Python math packages `numpy`, the scientific Python library `scipy`, and the math plotting library `matplotlib`. Let's install them:

```
sudo apt-get install python-numpy python-scipy python-matplotlib
python-sympy
```

I'll cover the other Python libraries we will use later (OpenCV, scikit-learn, Keras, etc.) as we need them in the appropriate chapters.

Setting up Nvidia Jetson Nano

For this setup, we will use an image to run Ubuntu 20.04 on our Jetson Nano, which is required for ROS 2. One source for this version is `https://github.com/Qengineering/Jetson-Nano-Ubuntu-20-image`.

The basic steps, which you can follow in the Git repo, are as follows:

1. The first step is to prepare an SD card with the operating system image on it. I used **Imager**, but there are several programs available that will do the job. You need an SD card with at least 32 GB of space – and keep in mind you are erasing the SD card in this process. This means that you need a card greater than 32 GB to start with – I used a 64 GB SD card as a 32 GB SD card did not work, contrary to the instructions provided on the website.

2. Follow the directions with your SD card – the Jetson Nano Ubuntu website (`https://github.com/jetsonhacks/installROS2`) advises us to use a Class 10 memory card with 64 GB of space. Put the SD card in your reader and start up your disk imager program. Double (and triple) check that you pick the right drive letter – you are erasing the disk in that drive. Select the disk image you downloaded. Hit the **Write** button and let the formatter create your disk image on the SD card:

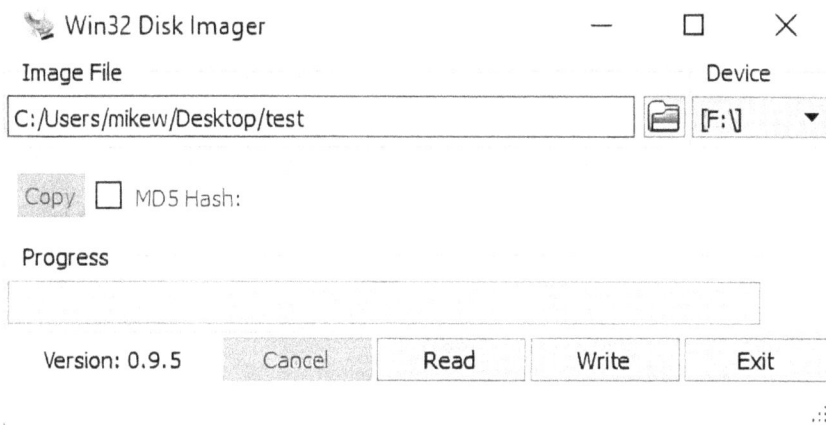

Figure 2.7 – Imager program to write out disk images on SD cards

3. You can follow the usual setup for setting up your language and keyboard, as well as setting up the network. I like to use a static IP address for the robot since we will be using it a lot.

4. It is always a good idea to set a new user ID and change the default passwords.

Now let's look at how we can install ROS 2.

Installing ROS 2

We need to install ROS 2 on the Jetson Nano. I used the *Foxy* version on my machine. You can follow the instructions at this link: `https://github.com/Razany98/ROS-2-installation-on-Jetson-Nano`.

You will have to set up the sources and point your computer at the ROS 2 repository. To do this, follow these steps:

1. Set `locale` using the following code:

    ```
    locale
    sudo apt update && sudo apt install locales
    sudo locale-gen en_US en_US.UTF-8
    sudo update-locale LC_ALL=en_US.UTF-8 LANG=en_US.UTF-8
    export LANG=en_US.UTF-8
    locale
    ```

2. Set up the source repository to use:

    ```
    apt-cache policy | grep universe or
    sudo apt install software-properties-common
    sudo add-apt-repository universe
    sudo apt update && sudo apt install curl gnupg2 lsb-release
    sudo curl -sSL https://raw.githubusercontent.com/ros/rosdistro/
    master/ros.key -o /usr/share/keyrings/ros-archive-keyring.gpg
    echo "deb [arch=$(dpkg --print-architecture) signed-by=/usr/
    share/keyrings/ros-archive-keyring.gpg] http://packages.ros.org/
    ros2/ubuntu $(source /etc/os-release && echo $UBUNTU_CODENAME)
    main" | sudo tee /etc/apt/sources.list.d/ros2.list > /dev/null
    ```

3. Install the ROS packages:

    ```
    sudo apt update
    sudo apt upgrade
    sudo apt install ros-foxy-desktop
    sudo apt install ros-foxy-ros-base
    ```

4. Set up the environment:

    ```
    source /opt/ros/foxy/setup.bash
    ```

5. Once you are done, you can check that your installation comepleted correctly by typing in the following:

    ```
    ros2 topic list
    ros2 node list
    ```

Before we proceed, let's take a look at how exactly ROS works.

Understanding how ROS works

You can think of ROS as a type of *middleware* that works to connect different programs together. It provides **Interprocess Communications (IPC)** between programs so we don't have to put all of our functions in one big block of code – we can distribute our robot's capabilities and develop them independently.

Each individual part of a ROS robot control system is called a **node**. A node is a single-purpose programming module. We will have nodes that collect camera images, perform object recognition, or control the robot arm. With ROS, we can isolate these functions and develop and test them independently.

The various nodes (programs) talk to one another via **topics**. A topic is generally used to publish data for other nodes to use. ROS 2 uses a publish/subscribe mode to move information around the robot. For example, our camera software node takes images and then publishes that information on a topic called /image_raw. This standard message type includes data about the image format, as well as the image itself. We also publish camera data on the /camera_info topic, using the sensor_msgs/ CameraInfo format, which is described in the following image:

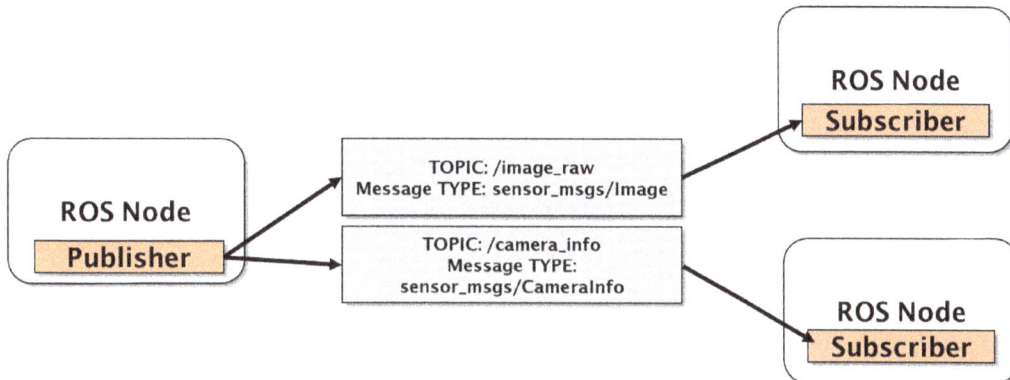

Figure 2.8 – ROS 2 nodes, topics, and message types

The /camera_info topic has a lot of valuable information about the image, or frame, including the timestamp the data was collected and the frame number. It also provides calibration information to help us understand the geometry of the captured image, which we can use to map pixels to the 3D space around the robot.

There is generally an existing or ROS standard message format for whatever you need to convey between components. I like to use the generic **string** (std_msgs/String) on a topic called RobotCmd to send general commands, such as mode changes, to the robot from a control application.

ROS 2 allows us to set **parameters** externally to the node (program). Each node can have its own list of parameters, and we can set parameters at any time. These are generally configuration commands that we set once and make our programs more generic and modular. For example, we may have a parameter for the robot arm that locks the rotation base, so that the arm can only be in front of the camera. We can create a parameter called `arm_base_lock`, define it as a Boolean, and use the following command:

```
ros2 param set /robot_arm arm_base_lock true
```

This will turn the rotation lock on. Then we can check this setting with this:

```
ros2 param get /robot_arm arm_base_lock
```

We get the following reply:

```
Boolean value is true
```

Since our robot will be composed of a number of nodes (programs) that all have to be started together, ROS 2 provides the concept of a **launch file** that lets us start all of our programs with one command. In ROS 1, launch files were built in **YAML** format. YAML stands for **Yet Another Markup Language**. In ROS 2 we can use YAML, Python, or **eXtensible Markup Language** (**XML**) to define a launch file. I'm used to creating files in YAML format, so we will stick to that. In our launch file, we can start nodes, change parameters, and create namespaces if we need to launch multiple copies of a node (for instance, if we had three cameras).

Virtual Network Computing

One tool that I have added to my Jetson Nano is **Virtual Network Computing** (**VNC**). This utility, if you are not familiar with it, allows you to see and work with the Nano desktop as if you were connected to it using a keyboard, a mouse, and a monitor. Since the Nano is physically installed inside the robot that travels by itself, attaching a keyboard, mouse, and monitor is not often convenient (or possible). There are many different versions of VNC, which is a standard protocol used amongst many Unix – and non-Unix – operating systems. The one I used is called **Vino**. You need two parts: the **server** and the **client**. The server runs on the Nano and basically copies all of the pixels appearing on the screen and sends them out to the Ethernet port. The client catches all of this data and displays it to you on another computer. Let's install the VNC server using the steps on this webpage: `https://developer.nvidia.com/embedded/learn/tutorials/vnc-setup`.

Load the viewer on your Windows PC, or Linux virtual machine, or do like I did, and load VNC on your Apple iPad. You will find the ability to log directly into the robot and use the desktop tools to be very helpful.

> **Important note**
>
> In order to get VNC to run on the Nano without a monitor attached, you must set the Nano to automatically log itself on. You can edit the `/etc/gdm3/custom.conf` file to enable automatic login:
>
> ```
> # Enabling automatic login
>
> AutomaticLoginEnable=true
>
> AutomaticLogin=[your username]
> ```

Setting up the colcon workspace

We will need a `colcon` workspace on your development machine—laptop or desktop—as well as on the Jetson Nano. Follow the instructions at `https://docs.ros.org/en/foxy/Tutorials/Beginner-Client-Libraries/Colcon-Tutorial.html`.

If you are already a user of ROS, then you know what a workspace is, and how it is used to create packages that can be used and deployed as a unit. We are going to keep all of our programs in a package we will call `albert`.

Summary

This chapter covered several important topics. It started with some of the basics of robotics, for readers who needed a bit more background. We talked about common robot parts, such as sensors, computers, and motors/actuators. We discussed the subsumption architecture in more depth and showed how it helps the robot arbitrate between responding to different events and commands. The next section covered the software setup for running the robot, including the offboard development environment and the onboard Jetson Nano computer environments. We set up the ROS and installed the Python tools.

The final section covered ROS 2 and explained what it is and what it does for us. ROS 2 is a middleware layer that lets us build modular components and multiple single-use programs, rather than having to lump everything into one executable. ROS also has logging, visualization, and debugging tools that help our task of designing a complex robot. ROS 2 is also a wonderful repository of additional capabilities that we can add, including sensor drivers, navigation functions, and controls.

In the next chapter, we will discuss how to go from a concept to a working plan for developing complex robot AI-based software using systems engineering practices such as use cases and storyboards.

Questions

1. Name three types of robot sensors.

2. What does the acronym PWM stand for?

3. What is analog-to-digital conversion? What goes in and what comes out?

4. Who invented the subsumption architecture?

5. Compare my diagram of the three-layer subsumption architecture to the Three Laws of Robotics postulated by Isaac Asimov. Is there a correlation? Why is there one, or why not?

   Hint: Think about how the laws change the behavior of the robot. Which is the lowest level law (from a subsumption perspective)? Which is the highest?

6. Do you think I should have given our robot project – *Albert* – a name? Do you name your robots? What about your washing machine? Why not?

7. What is the importance of the environment variable ROS_ROOT?

Further reading

- Scripts to install ROS 2 on Jetson Nano: `https://github.com/jetsonhacks/installROS2`

- Helpful troubleshooting in case you have problems with your ROS 2 installation can be found at `https://docs.ros.org/en/rolling/How-To-Guides/Installation-Troubleshooting.html`

- ROS 2 documentation: `https://docs.ros.org/en/foxy/index.html`

- Dr. Rodney Brooks's paper on the subsumption architecture: `https://people.csail.mit.edu/brooks/papers/AIM-864.pdf`

3

Conceptualizing the Practical Robot Design Process

This chapter represents a *bridge* between the preceding chapters on general theory, introduction, and setup, and the following chapters, where we will apply problem-solving methods that use **artificial intelligence** (**AI**) techniques to robotics. The first step is to clearly state our problem, from the perspective of the use of the robot, which is different from our view as the designer/builder of the robot. Then, we need to decide how to approach each of the hardware- and software-based challenges that we and the robot will attempt. By the end of this chapter, you will be able to understand the process of how to design a robot systematically.

This chapter will cover the following topics:

- A systems engineering-based approach to robotics
- Understanding our task – cleaning up the playroom
- How to state the problem with the help of use cases
- How to approach solving problems with storyboards
- Understanding the scope of our use case
- Identifying our hardware needs
- Breaking down our software needs
- Writing a specification

A systems engineering-based approach to robotics

When you set out to create a complex robot with AI-based software, you can't just jump in and start slinging code and throwing things together without some sort of game plan as to how the robot goes together and how all the parts communicate with one another. We will discuss a systematic approach to robot design based on **systems engineering** principles. We will be learning about use cases and will use storyboards as techniques to understand what we are building and what parts – hardware and software – are needed.

Understanding our task – cleaning up the playroom

We have already talked a bit about our main task for Albert, our example robot for this book, which is to clean up the playroom in my house after my grandchildren come to visit. We need to provide a more formal definition of our problem, and then turn that into a list of tasks for the robot to perform along with a plan of action on how we might accomplish those tasks.

Why are we doing this? Well, consider this quote by Steve Maraboli:

"If you don't know where you are going, how do you know when you get there?"

Figure 3.1 – It's important to know what your robot does

The internet and various robot websites are littered with dozens of robots that share one fatal character flaw: the robot and its software were designed first and then they went out to look for a job for it. In the robot business, this is called the **ready, fire, aim problem**. The task, the customer, the purpose, the use, and the job of the robot comes first. Another way of saying this is: to create an effective tool, the first step is to decide what you do with it.

I could have written this book as a set of theories and exercises that would have worked well in a classroom setting, which would have introduced you to a whole lot of new tools you would not know how to apply. However, this chapter is here to provide you with tools and methods to provide a path from having a good idea to having a good robot, with as little misdirection, pain, suffering, tears, and torn-out hair as possible.

> **Important note**
> You are on your own with burns; please be careful with the soldering iron.

The process we will use is straightforward:

1. The first step is to look at the robot from the user's perspective and then describe what it does. We will call these descriptions **use cases** – examples of how the robot will be used.

2. Next, we will break each use case down into **storyboards** (step-by-step illustrations), which can be word pictures or actual pictures. From the storyboards, we can extract tasks – a to-do list for our robot to accomplish.

3. The final step for this part of the process is to separate the to-do list into things we can do with software and things we will need hardware for. This will give us detailed information for designing our robot and its AI-based software. Keep in mind that one of the robot's uses is to be a good example for this book.

Let's start by looking at use cases.

Use cases

Let's begin our task with a statement of the problem.

Our robot's task – part 1

About once or twice a month, my five delightful, intelligent, and playful grandchildren come to visit me and my wife. Like most grandparents, we keep a box full of toys in our upstairs playroom for them to play with during their visits. The first thing they do upon arrival – at least the older grandkids– is take every single toy out of the toy box and start playing. This results in the scene shown in the following photograph – toys randomly and uniformly distributed throughout the playroom:

Figure 3.2 – The playroom in the aftermath of the grandchildren

Honestly, you could not get a better random distribution. They are really good at this. Since, as grandparents, our desire is to maximize the amount of time that our grandchildren have fun at our house and we want them to associate Granddad and Grandmother's house with having fun, we don't make them pick up the toys when they go home. You can see where this is heading.

By the way, if you are a parent, let me apologize to you in advance; this is indeed an evil plot on our, the grandparents, part, and you'll understand when you get grandkids of your own – and you will do this, too.

Where were we…? Yes, a room full of randomly and uniformly distributed foreign objects – toys – scattered about an otherwise serviceable playroom, which need to be removed. Normally, I'd just have to sigh heavily and pick up all this stuff myself, but I am a robot designer, so what I want to do is to make a robot that does the following:

1. Pick up the toys – and not the furniture, lights, books, speakers, or other items in the room that are not toys.

2. Put them in the toy box.

3. Continue to do this until there are no more toys to be found and then stop.

Here is a visual representation of this process:

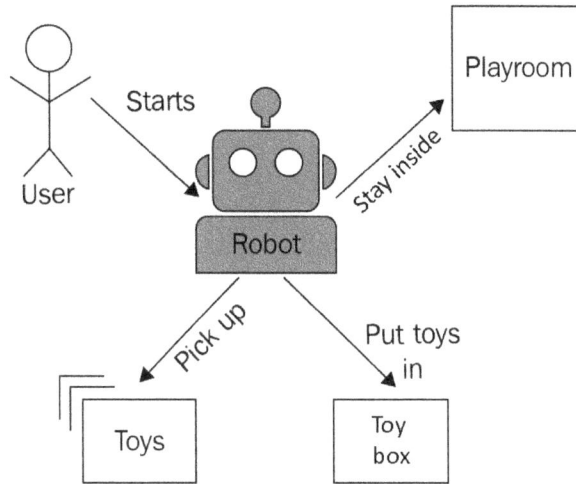

Figure 3.3 – Use case: pick up toys

Now we can ask some pertinent questions. I took journalism classes in school and I was taught the usefulness of the *5 Ws and an H – Who, What, When, Where, Why*, and *How*. These are just as useful for examining use cases. I've got one firm rule here in this section: no implementation details. Don't worry about how you are going to do this. Just worry about defining the results. So we'll leave out the *H* for now and focus on the *Ws*. Let's give this a try:

- **Who**: The robot. That was easy. We want the robot to do something, as in the robot does this and not me. What do we want the robot to do?

- **What**: This question can be answered in two ways:

  - **Pick up toys and put them in the toy box**: What does this answer tell us? It says we are going to be grasping and lifting something – toys. What are toys? We could also rephrase this as a negative, which brings us to the second answer.

  - **Pick up and put away in the toy box the items that were not previously in the room**: The toys were not in the room before the grandkids pulled them all out. So we either want to classify items as toys or as things that were not in the room before. *Not in the room* implies that the robot somehow knows what belongs in the room, possibly by making a survey prior to the children's arrival. However, *toys* implies that the robot can classify objects at least as *toys* and *not toys*. Let's stick with that for now. We may have some items in the room that are not toys but are out of place, and thus don't belong in the toy box. You can already see these questions shaping what comes later in this process.

- **When**: After the grandchildren have visited and they have left, continue to pick up toys until there are none left.

That gives us two conditions for *when*: a start and a stop. In this case, the start is defined as the grandkids have visited and they have left. Now, it is perfectly fair for me to state in the use case that I'll tell the robot when these conditions are met, since that is not putting me out. I'll be here, and I know when the room needs to be cleaned. Besides, I need to get the robot out and put it into the room. When not working, it stays on a bookshelf. So, let's change our *when* statement to the following:

When I (the user) tell you to, and don't stop until there are no more toys to be found.

Now, we could have decided that the robot needs to figure this out for itself and turn itself on after the grandchildren leave, but what is the return on investment for that? That would be a lot of work for not a lot of gain. The pain point for me, the user, is picking up toys, not deciding when to do it. This is a lot simpler.

Note that my *when* statement has a start and an end. Anyone who watched Mickey Mouse in the *Sorcerer's Apprentice* segment of *Fantasia* understands that when you have a robot, telling it when to stop can be important. Another important concept is defining the end condition. I did not say *stop when all of the toys are picked up* because that would imply the robot needed to know all of the toys, either by sight or number. It is easier as a task definition to say *stop when you see no more toys* instead, which accomplishes what we want without adding additional requirements to our robot.

It is perfectly normal to have to revisit use cases as the robot designer understands more about the problem – sometimes you can be working hard to solve a problem that is not relevant to solving the user's task. You can imagine some robot engineer in a team being given the task of *pick up all the toys* as meaning all toys ever invented, in all cultures, in all parts of the world! Then, you get a request for a $500,000 database software license and a server farm to house it. We just want to pick up the toys found in the playroom.

- **Where**: The playroom upstairs. Now we have some tricky parts. The area to be cleaned is a specific area of the house, but it is not really bound by walls. And it is upstairs – there is a stairway going down in the playroom that we don't want our robot tumbling down. How would you have known this? You won't unless you ask these kinds of questions! The environment the robot operates in is just as important as what it does. In this case, let's go back and ask the user. I'll stick in a floor plan for you here to define what I mean by *playroom*. On the bright side, we don't need to climb or descend stairs in this task. But we do need to look out for the staircase as a hazard:

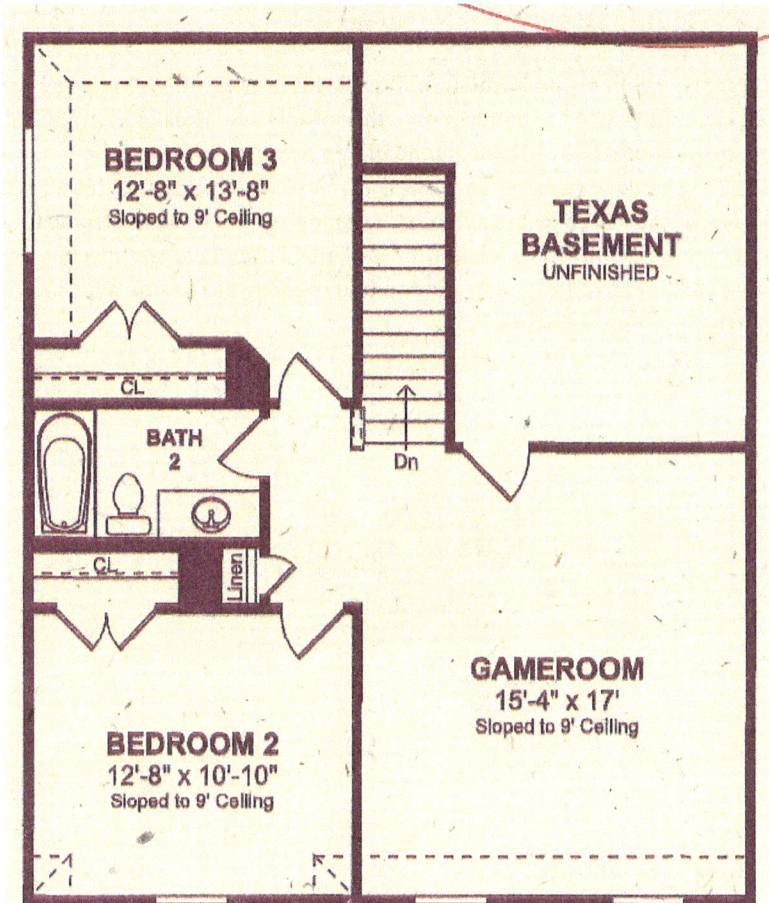

Figure 3.4 – The floor plan of my house, upstairs

- **Why**: So why is the robot picking up toys? I'm tempted to just write "Because *someone* has to do it." However, the answer is that I don't want the grandkids to pick up toys so that they have the maximum time to play, and I don't want to do it, either. So we are making a robot for this task. One maxim in the robot world is that proper tasks for robots are *dirty*, *dull*, or *dangerous*. This one definitely falls into the *dull* category.

Our robot has more than one use case – it has more than one function to perform.

Our robot's task – part 2

The robot needs to interact with my grandchildren. Why is this important? As I told you in *Chapter 1*, the grandchildren were introduced to some of my other robots, and the oldest grandkid, William, always tries to talk to the robots. I have three grandchildren who are on the autistic spectrum, so this is not an idle desire – I've read the research, such as *Robots for Autism* (`https://www.robokind. com/`), which states that robots can be helpful in such situations. While I'm not trying to do therapy, I'd like my robot to interact with my grandchildren verbally. I also have one specific request – the robot must be able to tell knock-knock jokes and respond to them, as they are a favorite of William. I want this robot to be verbally interactive.

So, here is a diagram of this use case:

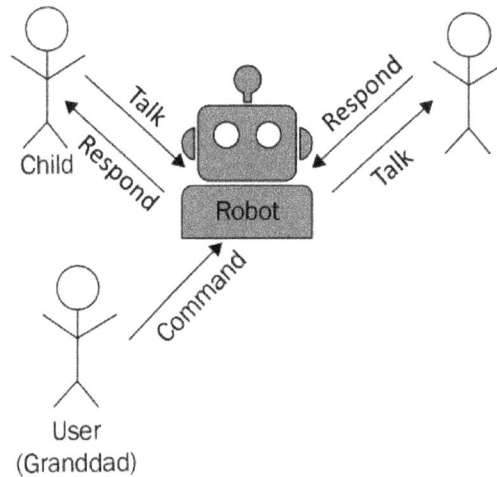

Figure 3.5 – Use case: interact with people

So let's go through the same exercise with this use case. We ask the pertinent questions: *who, what, when, where,* and *why*? Let's break these down:

- **Who**: The robot, the user (granddad), and the grandchildren.

  In this case, user interaction is part of the task. Who are we interacting with? I need to be able to command the robot to begin to interact. Then, we want the robot to both talk to and listen to the children.

- **What**: Receive commands and verbally interact (hold a conversation) with children, which must include knock-knock jokes. We keep the two kinds of functions: receive commands from – let's call me the **robot controller**, to make this more generic. The other function is to have a conversation with the children, including telling knock-knock jokes. We'll define *conversation* further on in our breakdown. You can refer to *Chapter 6* on using the robot as a **digital assistant**. We are going to use an open source digital assistant called *Mycroft* to act as a voice interface for the robot. We will add our own skills to the base Mycroft capability, which is actually quite versatile. The robot can get the weather, set timers, play music, look up information on Google (such as how many tablespoons in a quarter cup), and even tell you where the International Space Station is right now. But what it can't do is tell knock-knock jokes – until now, as we are adding this feature to the robot. Fortunately for us, the knock-knock joke has a very structured form based on puns and a call-and-response format that goes like this:

Robot: Knock knock.

Child: Who's there?

Robot: Lettuce.

Child: Lettuce who?

Robot: Lettuce (let us) in, we're freezing out here!

I'll leave diagramming the opposite form – responding to a knock-knock joke – to you.

- **When**: As requested by the robot controller, then when the child speaks to the robot.

  I think this is fairly self-explanatory: the robot interacts when sent a command to do so. It then waits for someone to talk to it. One thing we can extrapolate from this information is that when we are picking up toys, we are not expecting the robot to talk – the two activities are exclusive. We only pick up toys after the kids are gone, ergo there is one to talk to.

- **Where**: In the playroom, within about six feet of the robot.

  We have to set some limits on how far we can hear – there is a limit on how sensitive our microphone can be. I'm suggesting six feet as a maximum distance. We may revisit this later. When you come to a requirement like this, you can ask the customer *Why six feet?* They may say, *Well, that sounds like a reasonable distance.* You can then ask, *Well, if it was five feet, would that be a failure of this function?* And the user might respond, *No, but it would not be as comfortable.* You can continue to ask questions on distances until you get a feeling for the required distance (how far away to not fail), which might be three feet in this case (so that the child does not have to bend over the robot to be heard), and the desired distance, which is how far the user wants the function to work. These are important distinctions when we get around to testing. Where is the pass-fail line for this requirement?

- **Why**: Because my grandchildren want to talk to the robot, and have it respond (i.e., the users have specifically requested this feature).

Now, let's delve deeper into our robot's tasks.

What is our robot to do?

Now we are going to do some detailed analysis of what the robot needs to do by using the storyboard process. This works like this: We take each of our two tasks and break them down as completely as we can based on the answers to all of our *W* questions. Then we picturize each step. The pictures can be either a drawing or a word picture (a paragraph) describing what happens in that step. I like to start the decomposition process by describing the robot in terms of a **state machine**, which, for the first part of our problem, may be a good approach to understanding what is going on inside the robot at each step.

You are probably familiar with **state machine diagrams**, but just in case, a state machine diagram describes the robot's behavior as a series of discrete states or sets of conditions that define what actions are available to the robot:

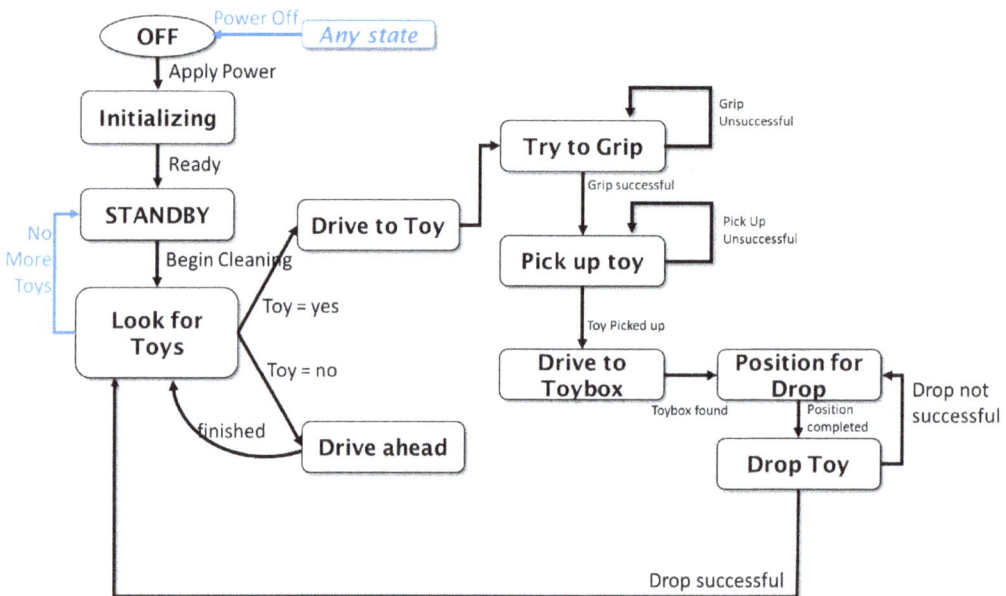

Figure 3.6 – Robot state machine diagram

Our first state is simply *Off* – the robot has no power turned on.

Each state is an event (or events) that causes the state to change. These are called **transitions**. To get from the *Off* state to whatever is next, some event has to occur – such as the human operator turning on the power. We'll call that transition event *Power applied*. Now what state are we in? There is some amount of time to get the computer booted and the programs loaded (*Initializing*). Once everything boots up and initializes, the robot will be ready to accept commands. Let's call this state *Standby*. The robot is just sitting waiting for instructions. Now we want to start cleaning the room. I send a *Begin cleaning* command to the robot, which changes the state to – what? What do we need to happen next? We could define a state called *Cleaning*, but that would encompass a lot of complex functions and we would not learn much from that. We need the robot to look for toys using its camera. If it does not find a toy, it needs to drive forward a short distance – avoiding obstacles – and then look again. In practice, we should be able to look for toys while driving without pausing constantly. We will need to make the *Look for toys* function interrupt driving when it sees a toy.

If it does find a toy, then the robot needs to position itself so that the toy is within reach of the robot arm. In the state machine diagram, we already added the transition *Begin cleaning*, which changes the state from *Standby* to *Look for toys*. Now we can add two more transitions: one called *Toy = no* and one called *Toy = yes*. The *Toy = no* branch goes to a state called *Drive ahead*, where the robot moves forward – while avoiding obstacles – and then goes back to the *Look for toys* state and tries again to find a toy. We will need some sort of means to tell the software how often to look for toys. We could use a simple timer – so many seconds elapsed. Or we could use some sort of distance function based on wheel motion.

So, now we have found a toy, what do we do? We need to drive to the toy, which puts it in range of our robot arm. We try to grip the toy with the robot's arm and hand. We may not be successful on the first try, in which case we want to try again. The loop transition, which is labeled *Grip unsuccessful*, says to go back and try again if you don't succeed the first time. Where have I heard that before? You can see the same with *Pick up toy*. Why are there two parts? We need to first get a hold of the toy before we can lift it. So I thought it needed two states, since we may fail to get a grip – the toy falls out of the hand, separately from picking the toy up, where the toy is too heavy or awkward to lift.

OK, we found a toy and picked it up. What is next? We need to put it in the toy box. The next state is *Drive to toy box*. Don't worry about *how* at this stage; this is just what we need to do. Later, we can further decompose this state into a more detailed version. We drive until we get to the event *Toy box found*. That means we see the toy box. Then we go to the *Position for drop* state, which moves the robot to a place where it can drop the toy in the box. The final state, *Drop toy*, is self-explanatory. We've dropped the toy, the robot has nothing in its gripper, and guess what? We start over by returning to the *Look for toys* state. If the robot decides that the drop was not successful (the toy is still in the gripper), then we have it try that step again, by repositioning the hand over the toy box and trying to drop the toy again by opening the hand. How do we know whether the gripper is empty? We try to close the grip and see what position the hand servo is in. If the gripper can close (go to a minimum state), then it is empty. If the toy falls outside the toy box (the robot misses the box entirely), then it is once again a toy on the floor, and will be treated in the normal manner – the robot will find it, pick it up, and try again.

This is all well and good, and our little robot goes around forever looking for toys, right? We've left out two important transitions. We need a *No more toys* event, and we need a way to get back to the *Off* state. Getting to *Off* is easy – the user turns off the power. I use the shorthand method of having a block labeled *Any state* since we can hit the off button at any time, no matter what else the robot is doing, and there is nothing the robot can or should do about it. It may be more proper to draw a line from each state back to *Off*, but that clutters the diagram, and this notation still gets the meaning across. The new state machine diagram looks like this:

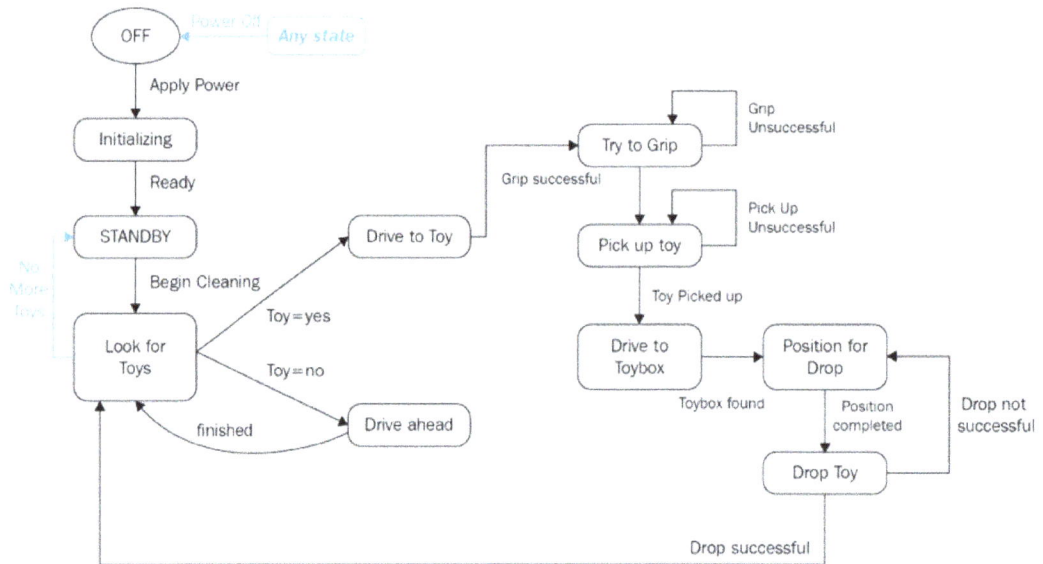

Figure 3.7 – New state machine diagram

Let's take a minute and talk about the concept of *No more toys*. How do we define this? This may take some experimentation, but for now, we'll say if we have not found a toy after 10 minutes of trying, then we are satisfied that there are no more toys to be found. Later, we can adjust that time as necessary. It is possible that 5 minutes is totally adequate for a room our size. Note that the *No more toys* event can only come from the *Look for toys* state, which should make sense.

We mentioned that the robot needs to avoid obstacles. But we don't have a state called *Avoid obstacles*. Why is that? That is because several of the states include driving, and each of those includes avoiding obstacles. It would not be appropriate to have a state for avoiding obstacles, since it is not unique to one state. What we need is a separate state machine that describes the robot's driving. As I mentioned in the *Introducing subsumption architecture* section in the last chapter, we can have more than one goal operational at a time.

The task of picking up toys is the mission, which is the overall goal of the robot. *Avoid obstacles* is a goal of the driving engine, the mid-level manager of our robot.

We've discussed our use cases and drawn a state machine diagram, so now let's move on to the next step, which is to create our storyboards.

Using storyboards

In this section, we are going to decompose our use cases further in order to understand the various tasks our robot must undertake on our behalf in the course of its two missions. I've created some **storyboards** – quick little drawings – to illustrate each point.

The concept of storyboards is borrowed from the movie industry, where a comic-strip-like narration is used to translate words on a page in the script into a series of pictures or cartoons that convey additional information not found in the script, such as framing, context, movement, props, sets, and camera moves. The practice of storyboarding goes all the way back to silent movies and is still used today.

We can use storyboards in robotics design for the same reasons: to convey additional information not found in the words of the use cases. Storyboards should be simple, quick, and just convey enough information to help you understand what is going on.

Let's get started. We are not going to create storyboards for *Power applied*, *Initializing*, or *Standby* because a storyboard is not really needed for those simple concepts. We will jump ahead to the *Begin cleaning* event in our state diagram.

Storyboard – put away the toys

When our story begins, what is the robot doing? It has been turned on, and is in a standby state waiting to be told what to do. How does it receive a command? A nice, hands-free way would be to receive a voice command to *begin cleaning*, or some similar words that mean the same thing.

The next step in our process after *Begin cleaning* is *Look for toys*. This storyboard frame is *what the robot sees* as it is commanded to start cleaning. It sees the room, which has three kinds of objects visible – that is, toys, things that are not toys (the ottoman and the fireplace), and the room itself, including the walls, and the floor:

Figure 3.8 – Waiting for a voice command to begin cleaning

We could select any sort of sensor to detect our toys and direct our robot. We could have a LiDAR, thermal, or sonar scanner. Let's hypothesize that the best sensor tool for this task is a regular USB camera. We have control of the lighting, the toys are not particularly warmer or cooler than the surroundings, and we need enough information to identify objects by type. So, video it is. We will determine later exactly what kind of camera we need, so add that to our *to-do list*.

2

Not toy

Toy

Not toy

Toy

Toy

Toy

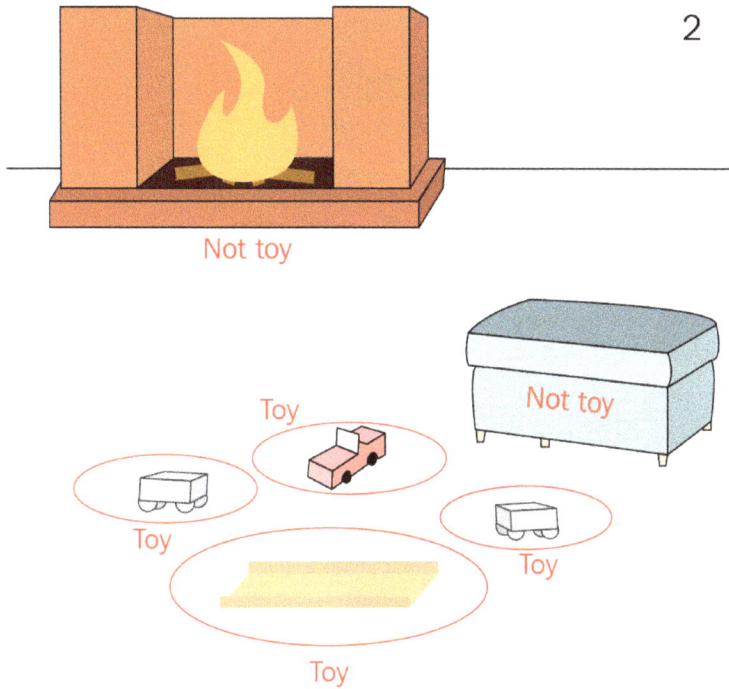

Figure 3.9 – Look for toys

Our next storyboard is to *look for toys*. We need to run some sort of algorithm or technique to classify objects by type. The results of that algorithm are to find the objects – separate them from the background of the floor – and then classify each object as a toy or not a toy. We don't really care to have any more breakdown than that – we leave all *Not toy* objects alone, and pick up all *Toy* objects. Note that we draw circles around the objects that are toys, which is another way of saying that we must locate them in the camera frame.

So what does this simple picture tell us we did not know before? It tells us the following:

- We need to segment the camera image by objects

- We need to locate the objects in the camera frame

- We need to classify the objects as either *Toy* or *Not toy*

- We need to be able to store and remember this information

We only can pick up and move one toy at a time – we only have one hand, and nobody said in the use cases that we need to pick up more than one at a time. So, we only care about one toy – and let's arbitrarily say we pick up the closest one to the robot:

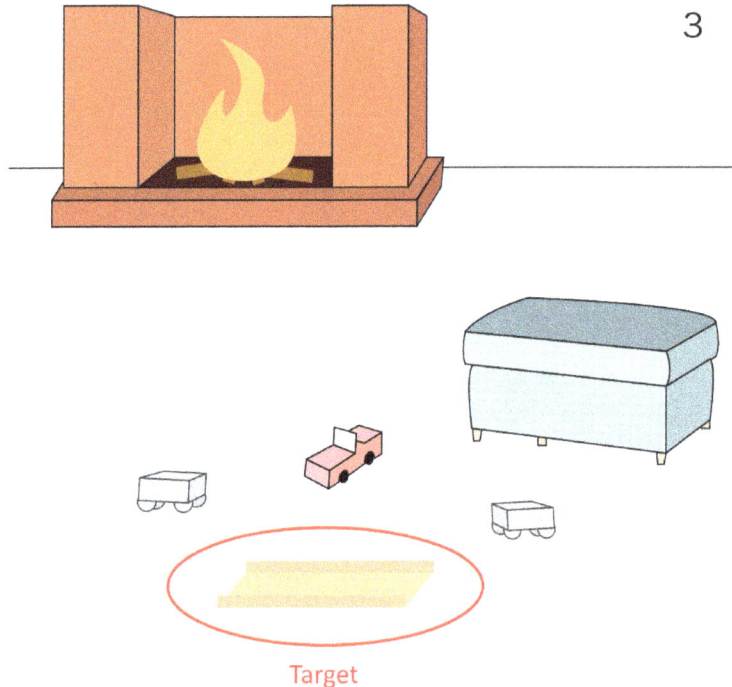

Figure 3.10 – Select nearest toy

We might also say that it's the toy that is easiest to get to, which might be a slightly different process than choosing the closest one. We set that toy to be the target for our next action, which is what? If you said to drive to the toy, you would be correct. However, we must not just drive to the toy but put the robot's body in a position to use the robot arm to grasp the toy. By the way, that means the robot arm must be able to reach the ground or very close to the ground, as we have some small toys.

Our robot must plan a route from its current position to a spot where it can attempt to pick up the toy. We set a target goal an arms-length away from the center of the toy:

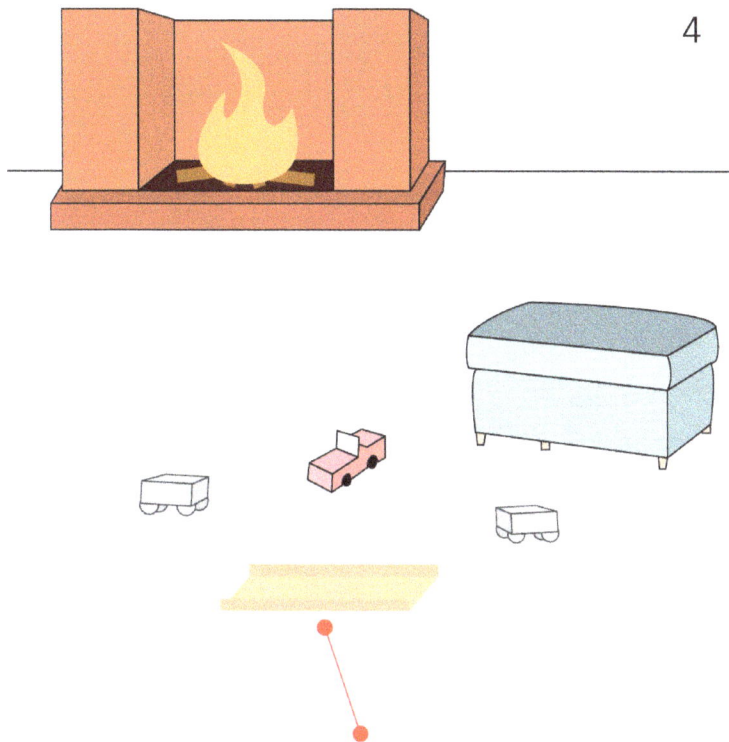

Figure 3.11 – Plan route to target

The robot needs to make sure that there are no obstacles en route. There are two ways of doing this. As illustrated, we can clear the path that the robot is traveling on by adding the width of the robot (plus a bit of extra) and see whether any obstacles are in that area, or we can add a border around obstacles and see whether our path goes into those boundaries. Regardless, we need to have a path free of obstacles:

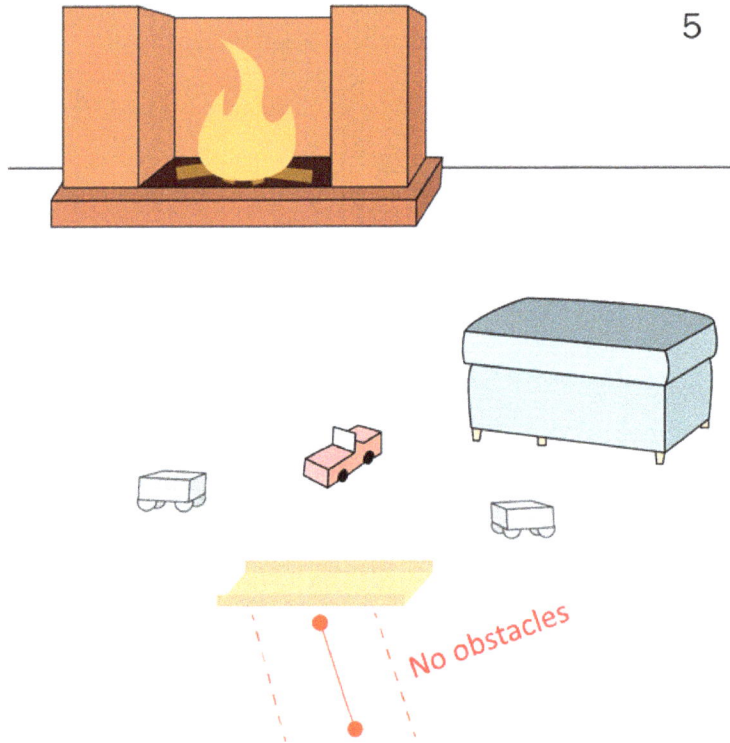

Figure 3.12 – Look for obstacles on the route

The robot determines for itself the proper alignment to prepare to pick up the toy:

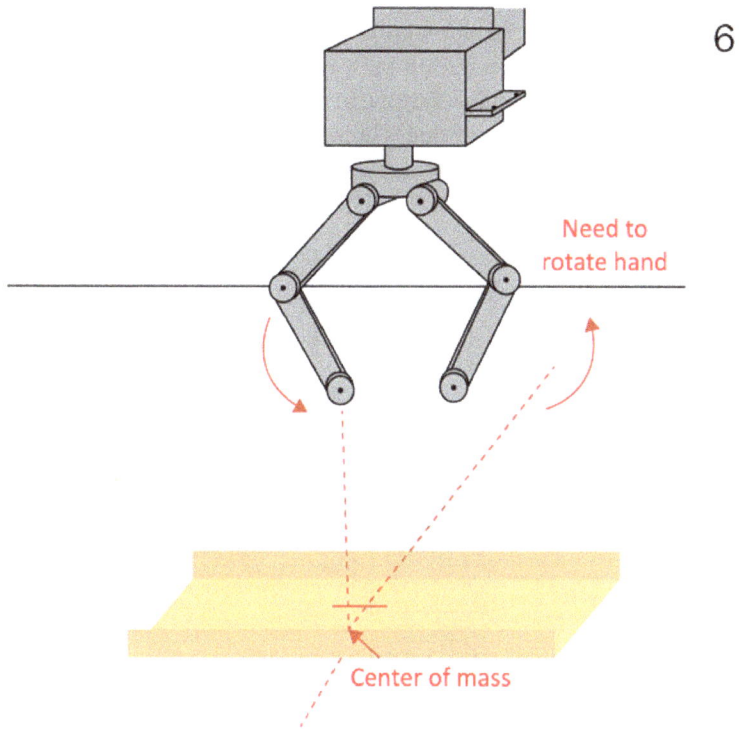

Figure 3.13 – Position robot hand

Now that the robot has completed its drive, the robot can move the robot hand to a position to pick up the toy. We need to put the robot hand over the center of mass of the toy, and then rotate the hand to match a narrow part of the toy so we can pick it up. One of our goals for this project is to not dictate how the robot does this, but rather to let it learn for itself. So, we can say for this storyboard panel that the robot uses its training and machine learning to use an appropriate hand pose to prepare to grasp the object. We can surmise that that includes lining the hand up:

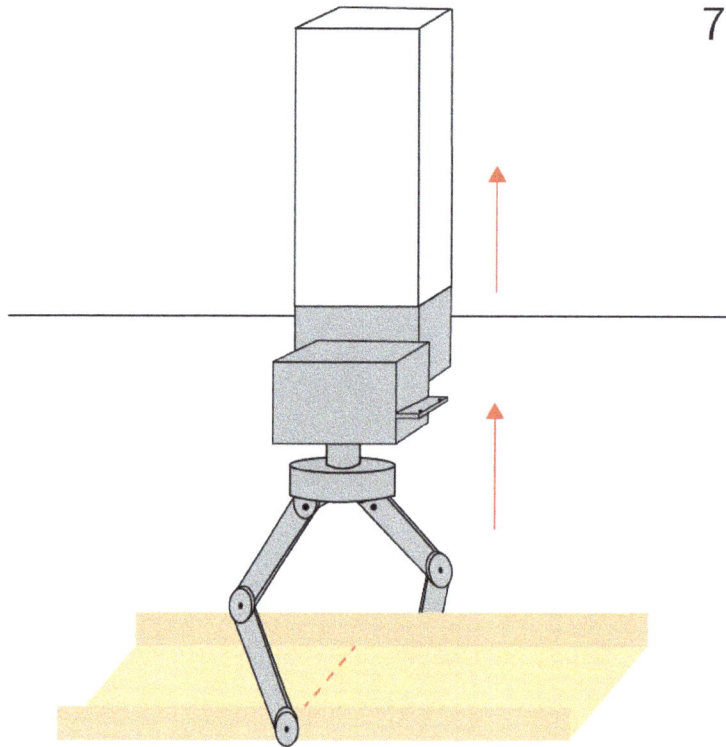

Figure 3.14 – Pick up toy

Probably *storyboard 6* is the hard part (*Figure 3.13*), and in *storyboard 7*, the robot completes the grasp of the object and picks it up (*Figure 3.14*). The robot has to be able to determine whether the pick-up was successful, and if not, try again. That was in the state machine diagram we did before. We have now picked up the toy. What's next? Find the toy box!

8

Toy box

Obstacle

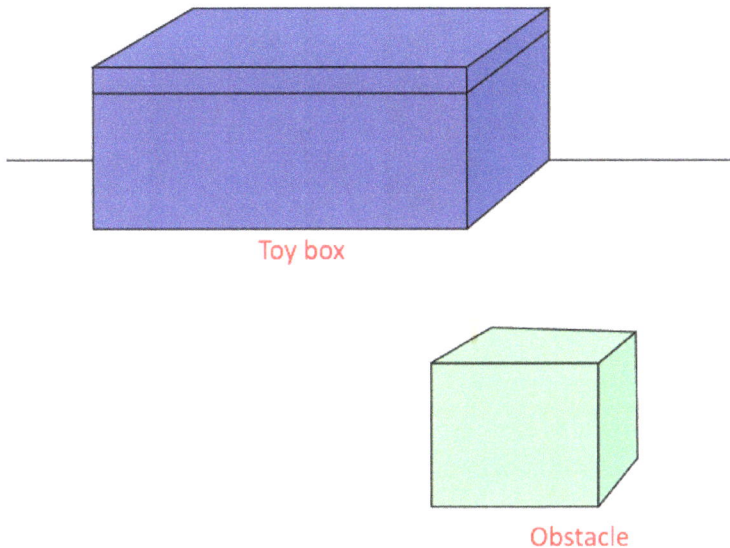

Figure 3.15 – Find toy box

Now we need the robot to find the toy box. Again, we don't care how at this point. We are still worried about *what* and not *how*. Somehow, the robot looks around and finds the toy box, which, in this case, is large, against the wall, and has a distinctive color. Regardless, the robot has to find the toy box on its own. The labels in the picture indicate that the robot can distinguish the toy box and that it considers all other objects it perceives as obstacles. We can see we don't need to have the *Toy/Not toy* capability active at the same time, only the *Toy box/Not toy box* decision-making process. This does reduce some of the required processing and will make machine learning easier.

Now that we have found the toy box, we illustrate a slightly more complex task of navigating around an obstacle to get there. In this example, we show the purple outline of the robot's base, compared to a red outline around the obstacle, which I labeled *Keep out zone*. This gives us more guidance on how to avoid obstacles:

9

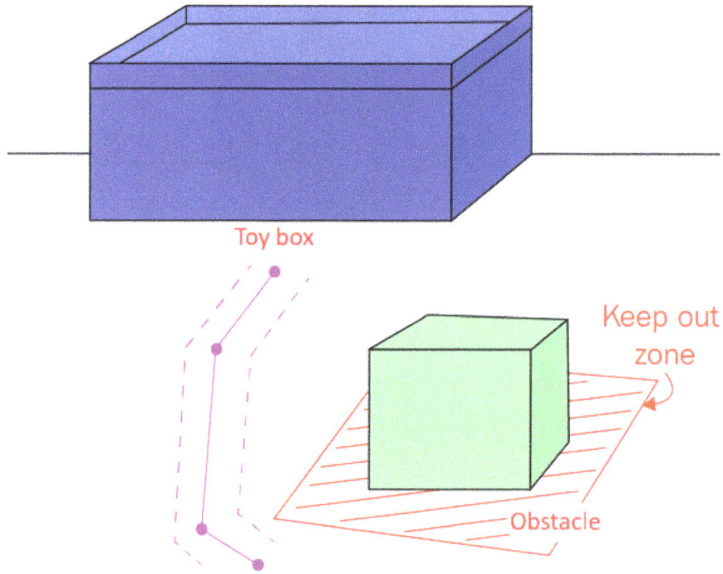

Figure 3.16 – Plan path to toy box

We want to keep the center of the robot out of the *Keep out zone*. We need to get close enough to the toy box to drop our toy into it:

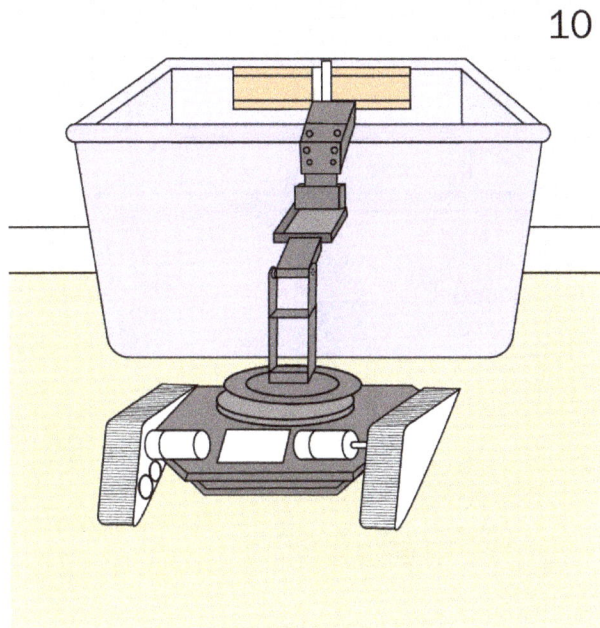

Figure 3.17 – Align toy with box

In *storyboard 10*, we lift the toy high above the top of the toy box and position our toy to fall inside the toy box when we let go of it. Make a note that we have to have the toy lifted before the final few inches to the toy box. We put the robot hand over the top of the opening of the toy box, just as far forward as we can and in the middle of the toy box.

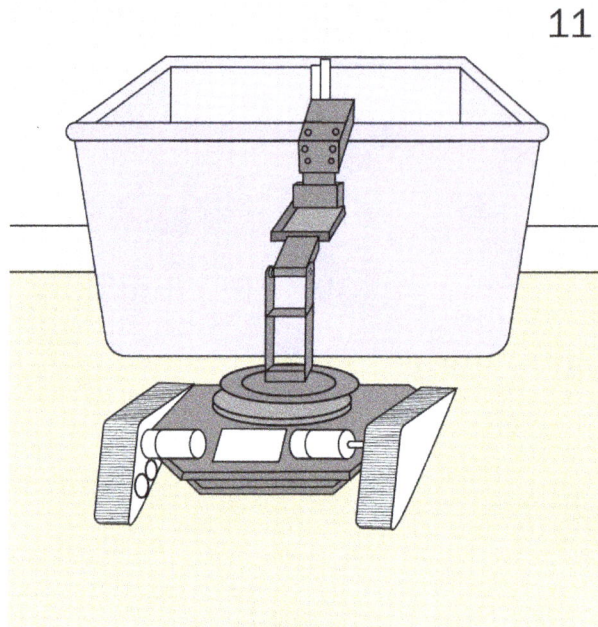

Figure 3.18 – Drop toy in box

Our final step in the toy saga is to open the robot hand and let the toy hopefully fall into the toy box. I predict that we will have to spend some trial and error time getting this right. We may have to tilt the open hand right and left to get the toy to drop. If the toy falls outside of the box, then it is not put away and we have to start all over and try to put it away again. We don't need a new state for this because it returns to being a toy on the floor, and we already have a state for that.

I hope that you have seen in the storyboard process how this provides insight into visualizing the robot's tasks. I would say the most important benefit is that it forces you to think about what the robot is doing and to break down each step into smaller and smaller parts. Don't hesitate to take this storyboard and break an individual panel down into its own storyboard, if that is what you feel you need to do.

Project goals

Since this is an AI/machine learning project, we must add to our project goals not just putting away toys but also using machine learning, adaptive systems, neural networks, and other tools to provide a new approach to solving these sorts of problems. You may think, "*Why bother? You can do this better with a standard programming approach.*" I would say from experience that these problems are difficult to solve that way, and you can do your own research to see where companies, large and small, have tried to solve this sort of problem and failed – or at least not succeeded. This problem is not easily solved by any means, and using an AI-based approach has a far greater chance of success than standard programming techniques. Now, I'm not saying we are going to succeed beyond our wildest dreams at this task in this book, but our objective is to learn a whole lot along the way!

So, we pause at this point in defining our project to say that we are deliberately choosing to use artificial intelligence and machine learning as an approach to solving a problem that has proven to be difficult with other means.

Since we are going to be teaching the robot various tasks, it will be more effective if we can teleoperate the robot and drive it around like a radio-controlled car, in order to collect data and take pictures that we will use for object recognition later. We don't need this for operations, we need this for training. We will add this required operation to our *to-do list*.

In our next step, we are going to extract from all of our hard work the **hardware** and **software** tasks that our robot will have to accomplish. But before we do this, let's pause for a moment to discuss a common mistake made in defining the scope of the use case.

Understanding the scope of our use case

Desirements (a word made up by combining *desire* and *requirements*) are functions that would be *nice to have* but not strictly necessary. For example, if we decided to add flashing lights to the robot because it looks cool, that would be a desirement. You may want to have it, but it does not contribute to the mission of the robot or the task it needs to perform.

Another example would be if we added that the robot must operate in the dark. There is no reason for this in the current context, and nothing we've stated in the use cases said that the robot would operate in the dark – just in an indoor room. This would be an example of **scope creep**, or extending the operation conditions without a solid reason why. It's important to work very hard to keep requirements and use cases to a minimum, and even to throw out use cases that are unnecessary or redundant. I might have added a requirement for sorting the toys by color, but sorting does not help with picking up the toys, and besides, I've only got one toy box. I might have added the task in the interest of education for you, my readers, but it does not help with that objective either, so color sorting is not included.

Now, let's proceed to identifying our hardware requirements.

Identifying our hardware needs

Based on our storyboards, I extracted or derived the following hardware tasks:

- Drive the robot base

- Carry the robot arm

- Lift toys

- Put toys in the toy box (arm length)

- Sensors:

  - Arm location

  - Hand status (open/close)

  - Robot vision (camera) for obstacle avoidance

- Provide power for all systems:

  - 5V for Nvidia Nano

  - 5V for Arduino

  - Arm power – 7.2V

  - Motor power – 7.2V

- Onboard computers:

  - A computer that can receive commands remotely (Wi-Fi Nano):

    - Runs ROS 2

    - Runs Python 3

  - A computer that can interface with a camera

  - A computer that can control motors (Arduino)

  - An interface that can drive servo motors for the robot arm (servo controller)

Now, let's take a look at the software requirements.

Breaking down our software needs

This list of software tasks was composed by reviewing the state machine diagram, the use cases, and the storyboards. I've highlighted the steps that will require AI and will be covered in detail in the coming chapters:

1. **Power on self-test (POST):**

   I. Start up robot programs.

   II. Check that the Nano can talk to the Arduino and back.

   III. Try to establish communications with the control station.

   IV. Report POST success or failure as appropriate and enter in the log.

2. Receive commands via Wi-Fi for teleoperation:

   - Drive motor base (right/left/forward/back)

   - Move hand up/down/right/left/in/out/twist

   - Record video or record pictures as image files

3. Send telemetry via Wi-Fi.

4. Monitor progress.

5. Send video.

6. Navigate safely:

   - Learn to avoid obstacles

   - Learn to not fall down stairs

7. Find toys:

   - Detect objects

   - Learn to classify objects (Toy/Not toy)

   - Determine which toy is closest

8. Pick up toys:

    I.   Move to the position where the arm can reach the toy

    II.   Devise a strategy for grasp

    III.   Attempt grasp

    IV.   Determine whether grasping was successful

    V.   If not, try again with a different strategy

    VI.   Reweight grasp technique score based on success

9. Put toys in the toy box:

    - Learn to identify the toy box

    - Find the toy box

    - Drive to the known toy box location using navigation

    - Move to the dump location:

        - Avoid obstacles

        - Lift the toy above the toy box lid

    - Drop the toy

    - Check to see whether the toy drop was successful

    - If not, reposition and try again

    - If the toy misses the toy box, we treat it as a toy on the floor again

10. Determine there are no more toys.

11. Stand by for instructions.

12. Teleoperate:

    - Move base forward/backward/left/right

    - Move arm up/down/right/left

    - Move hand in/out/twist/open/close

    - Record video/take pictures

13. Simulate personality:

    - Talk

    - Listen/recognize words

    - Understand some commands

    - Tell knock-knock jokes

    - Understand knock-knock jokes

14. Voice commands:

    - Clean-up room

    - Put this away

    - Come here

    - Stop

    - Wait

    - Resume

    - Go home

    - Turn left/ right

    - Forward/ back

    - Hand up/hand down

    - Hand left/hand right

    - Open hand/close hand

In this list, where did I get *teleoperate*? We don't remember discussing that in the use cases and storyboards. We are going to need to teach the robot to navigate and find toys, and for that, we need to move the robot around and take pictures. One easy way to do that is by driving the robot around with **teleoperations** (remote control).

Writing a specification

Our next task is to write specifications for our various components. I'll go through an example here that we must do as part of our toy-grasping robot project: we need to *select a camera*. Just any old camera will not do – we need one that meets our needs. But what are those needs? We need to write a camera specification so that when we are looking at cameras to buy, we can tell which one will do the job.

We've created our storyboard and our use cases, so we have the information we need to figure out what our camera needs to do. We can reverse engineer this process somewhat: let's discuss what things make one camera different from another. First of all is the interface: this camera goes on board the robot, so it has to interface with the robot's computer, which has USB, Ethernet, and a special camera bus. What other things about cameras do we care about? We certainly care about cost. We don't want (or need) to use a $1,000 camera for our inexpensive robot. Cameras have resolution: the number of pixels in each image. That can vary from 320 x 240 to 4,000 x 2,000 (4K). Cameras also have a field of view, which is the number of angular degrees the camera can see. This can vary from 2.5 degrees (very narrow) to 180 degrees (very wide). There are also cameras that see in the dark or have various types of infrared sensitivity. Finally, there is size and weight; we need a small camera that fits on our robot.

This makes the parameters that we need to decide the following:

- **Field of view**: [180 - > 2.5]
- **Resolution**: [320 x 280 -> 4,000 x 2,000]
- **Cost**: (low to high) – cheaper is better
- **Sees in the dark**: Yes/no
- **Size and weight**: Smaller and lighter is much better; must fit on the robot
- **Interface**: USB, Ethernet, or camera bus; power >11V

The reason for listing these parameters like this is that we can now concentrate on those features that we can select, so we are not wasting time looking at other parameters that we don't care about. Let's see whether we can knock off some of the parameters:

- If we use USB as the **interface**, the power is provided by the connector, and we don't need extra cables or routers. This is also the lowest cost method, so we choose USB as the interface.
- We also don't have any requirements in our use cases to **see in the dark**, so we don't need a special infrared camera.
- The next question is to determine the **field of view**. We need to see the entire area where the robot arm can move in as it picks up a toy. We also need enough field of view to see when we are driving to avoid obstacles. We can take some measurements from the robot, but we can quickly see that we mostly need to see close to the robot, and we can't see past the tracks on either side. This sets the field of view required to be close to 90 degrees. More field of view than this is acceptable, less is not.

- Our final problem is determining the **resolution** we need to perform our object recognition. For that, we need an additional data point – how many pixels do we need to recognize an object as a toy? That is what we will do with this camera – recognize toys and things that are not toys. We also have to pick a distance at which we can recognize the toy. We don't have a firm requirement out of the use cases, so we have to make an educated guess. We know that our room is 17 feet long, and it has furniture in it. Let's guess that we need 8 feet of distance. How do we know this is correct? We do a thought experiment. If we can identify a toy 8 feet away, can we accomplish our task? We can see the toy half a room away. That gives the robot plenty of space to go drive to the toy and it won't spend much time looking for toys. As a check, if the robot had to be 4 feet away to recognize a toy, would that be unusable? The answer is probably not – the robot would work OK. How about 3 feet? Now we are getting to the point where the robot has to drive right up to the toy to determine what it is, and that might result in more complicated logic to examine toys. So, we say that 3 feet is not enough, 4 feet is acceptable, and 8 feet would be great.

  What resolution is required in the camera to recognize a toy at 8 feet with a 90-degree lens? I can tell you that the ImageNet database requires a sample 35 pixels wide to recognize an object, so we can use that as a benchmark. We assume at this point that we need an image at least 35 pixels across. Let's start with a camera with *1,024 x 768* pixels, which is 1,024 pixels wide. We divide by 90 degrees to get that each degree has 11.3 pixels (*1,024/90*). How big is our smallest toy at 8 feet? Our smallest toy is a Hot Wheels toy, which is approximately 3 inches long. At 8 feet, this is 1.79 degrees or 20.23 pixels (*1.79 degrees x 11.3 pixels/degree*). That is not enough. Solving the distance equation for 3 inches, we get a maximum distance of 4.77 feet for a camera with *1,024 x 768* pixels. That is just barely acceptable. What if we had an HD sensor with *1,900 x 1200* pixels? Then, at 8 feet, I get 75 pixels – more than enough to give us the best possible distance. If we use a sensor 1,200 pixels wide, we have a recognition distance of 5.46 feet, which is adequate but not great.

I walked you through this process to show you how to write a specification and the types of questions you should be asking yourself as you decide what sensors to acquire for your project.

Summary

This chapter outlined a suggested process for developing your to-do list as you develop your robot project. This process is called systems engineering. Our first step was to create use cases or descriptions of how the robot is to behave from a user's perspective. Then, we created more detail behind the use cases by creating storyboards, where we went step by step through the use case. Our example followed the robot finding and recognizing toys, before picking them up and putting them in the toy box. We extracted our hardware and software needs, creating a to-do list of what the robot will be able to do. Finally, we wrote a specification for one of our critical sensors: the camera.

In the next chapter, we will dive into our first robot task – teaching the robot to recognize toys using computer vision and neural networks.

Questions

1. Describe some of the differences between a storyboard for a movie or cartoon and a storyboard for a software program.

2. What are the five *W* questions? Can you think of any more questions that would be relevant to examine a use case?

3. Complete this sentence: A use case shows what the robot does but not _____.

4. Take *storyboard 9* in *Figure 3.16*, where the robot is driving to the toy box, and break it down into more sequenced steps in your own storyboard. Think about all that must happen between *frames 9 and 10*.

5. Complete the reply form of the knock-knock joke, where the robot answers the user telling the joke. What do you think is the last step?

6. Look at the teleoperate operations. Would you add any more, or does this look like a good list?

7. Write a specification for a sensor that uses distance measurement to prevent the robot from driving downstairs.

8. What is the distance at which a camera with 320 x 200 pixels and a 30-degree field of view can see a 6-inch wide stuffed animal, still assuming we need 35 pixels for recognition?

Further reading

For more information on the topics in this chapter, you can refer to the following resources:

- *A Practical Guide to SysML: The Systems Modeling Language*, by Sanford Friedenthal, Alan Moore, and Rick Steiner, published by Morgan Kaufman; this is the standard introduction to **Model-Based Systems Engineering** (**MBSE**)

- *The Agile Developer's Handbook* by Paul Flewelling, published by Packt

Part 2:
Adding Perception, Learning, and Interaction to Robotics

To see, understand, and interact with the environment, robots need to have perception. AI is one approach that can be used for recognizing objects and navigation. This part empowers you with the essential skills to efficiently operate your robots using AI techniques. Our example in this book is creating a robot that picks up toys, so we start with recognizing toys with a **neural network**. Then we work with the robot arm to pick up toys using tools such as **reinforcement learning** and **genetic algorithms**. The next chapter covers the creation of a robot digital assistant that can listen and understand your commands, and even tell knock-knock jokes.

This part has the following chapters:

- *Chapter 4, Recognizing Objects Using Neural Networks and Supervised Learning*
- *Chapter 5, Picking Up and Putting Away Toys Using Reinforcement Learning and Genetic Algorithms*
- *Chapter 6, Teaching the Robot to Listen*

4

Recognizing Objects Using Neural Networks and Supervised Learning

This is the chapter where we'll start to combine **robotics** and **artificial intelligence** (**AI**) to accomplish some of the tasks we laid out so carefully in previous chapters. The subject of this chapter is **object recognition** – we will be teaching the robot to recognize what a toy is so that it can then decide what to pick up and what to leave alone. We will be using **convolutional neural networks** (**CNNs**) as machine learning tools for separating objects in images, recognizing them, and locating them in the camera frame so that the robot can then locate them. More specifically, we'll be using images to recognize objects. We'll be taking a picture and then looking to see whether the computer recognizes specific types of objects in those pictures. We won't be recognizing objects themselves, but rather images or pictures of objects. We'll also be putting bounding boxes around objects, separating them from other objects and background pixels.

In this chapter, we will cover the following topics:

- A brief overview of image processing
- Understanding our object recognition task
- Image manipulation
- Using YOLOv8 – an object recognition model

Technical requirements

You will be able to accomplish all of this chapter's tasks without a robot if yours cannot walk yet. We will, however, get better results if the camera is in the proper position on the robot. If you don't have a robot, you can still do all of these tasks with a laptop and a USB camera.

Overall, here's the hardware and software that you will need to complete the tasks in this chapter:

- Hardware:

  - A laptop computer

  - Nvidia Jetson Nano

  - USB camera

- Software:

  - Python 3

  - OpenCV2

  - TensorFlow

  - YOLOv8, which is available at `https://github.com/ultralytics/ultralytics`

The source code for this chapter can be found at `https://github.com/PacktPublishing/Artificial-Intelligence-for-Robotics-2e`.

In the next section, we will discuss what image processing is.

A brief overview of image processing

Most of you will be very familiar with computer images, formats, pixel depths, and maybe even convolutions. We will be discussing these concepts in the following sections; if you already know this, skip ahead. If this is new territory, read carefully, because everything we'll do after is based on this information.

Images are stored in a computer as a two-dimensional array of **pixels** or picture elements. Each pixel is a tiny dot. Thousands or millions of tiny dots make up each image. Each pixel is a number or series of numbers that describe its color. If the image is only a grayscale or black-and-white image, then each pixel is represented by a single number that corresponds to how dark or light the tiny dot is. This is straightforward so far.

If the image is a color picture, then each dot has three numbers that are combined to make its color. Usually, these numbers are the intensity of **Red, Green, and Blue (RGB)** colors. The combination (0,0,0) represents black (or the absence of all colors), while (255,255,255) is white (the sum of all colors). This process is called the additive color model. If you work with watercolors instead of computer pixels, you'll know that adding all the colors in your watercolor box makes black – that is a subtractive color model. Red, green, and blue are primary colors that can be used to make all of the other colors. Since an RGB pixel is represented by three colors, the actual image is a three-dimensional array rather than a two-dimensional one since each pixel has three numbers, making an array of (height, width, 3). So, a picture that is 800 x 600 pixels would be represented by an array of dimensions given by (800,600,3), or 1,440,000. That is a lot of numbers. We will be working very hard to minimize the number of pixels we are processing at any given time.

While RGB is one set of three numbers that can describe a pixel, there are other ways of describing the **color formula** that have various usages. We don't have to use RGB – for instance, we can also use **Cyan, Yellow, and Magenta** (**CYM**), which are the complementary colors to RGB, as shown in *Figure 4.2*. We can also break down colors using the **Hue, Saturation, and Value** (**HSV**) model, which classifies color by hue (shade of color), saturation (intensity of color), and value (brightness of color). HSV is a very useful color space for certain calculations, such as converting a color image into grayscale (black and white). To turn RGB into a grayscale pixel, you have to do a bit of math – you can't just pull out one channel and keep it. The formula for RGB to grayscale, as defined by the **National Television System Committee** (**NTSC**), is as follows:

$$0.299*Red + 0.587*Green + 0.114*Blue$$

This is because the different wavelengths of light behave differently in our eyes, which are more sensitive to green. If you have color in the HSV color model, then creating a grayscale image involves considering *V* (value) and throwing the *H* and *S* values away. As you can imagine, this is a lot simpler. This is important to understand as we will be doing quite a bit of image manipulation throughout this chapter. But first, in the following section, we'll discuss the image recognition task we will be performing in this chapter.

Understanding our object recognition task

Having a computer or robot recognize an image of a toy is not as simple as taking two pictures and then saying `if picture A = picture B, then toy`. We are going to have to do quite a bit of work to be able to recognize a variety of objects that are randomly rotated, strewn about, and at various distances. We could write a program to recognize simple shapes – hexagons, for instance, or simple color blobs – but nothing as complex as a toy stuffed dog. Writing a program that did some sort of analysis of an image and computed the pixels, colors, distributions, and ranges of every possible permutation would be extremely difficult, and the result would be very fragile – it would fail at the slightest change in lighting or color.

Speaking from experience, I had a recent misadventure with a large robot that used a traditional computer vision system to find its battery charger station. That robot mistook an old, faded soft drink machine for its charger – let's just say that I had to go buy more fuses.

What we will do instead is teach the robot to recognize a set of images corresponding to toys that we will take from various angles. We will do this by using a special type of **artificial neural network** (**ANN**) that performs convolution operations on images. It is classified as an AI technique because instead of programming our software to recognize objects by writing code, we will be training a neural network to correctly *segment* (separate from the rest of the image) and *label* (classify) groups of pixels in an image by how closely they resemble groups of labeled pixels that the network was trained on. Rather than the code determining the robot's behavior, it is the data we train the network on that does the work. Since we (the humans) will be training the neural network by providing segmented and labeled images, this is called **supervised learning**. This involves telling the network what we want it to learn and reinforcing (rewarding) the network based on how well it performs. We'll discuss **unsupervised learning** in *Chapter 8*. This process entails us not telling the software exactly what to learn, which means it must determine that for itself.

To clarify, in this section, we will tell the ANN what we want it to learn, which in this case is to recognize a class of objects we will call *toys*, and to draw a bounding box around those objects. This bounding box will tell other parts of the robot that a toy is visible and where it is in the image.

I'll be emphasizing the unique components we will use to accomplish our task of recognizing toys in an image. Do you remember what the storyboard from *Chapter 3* told us to do?

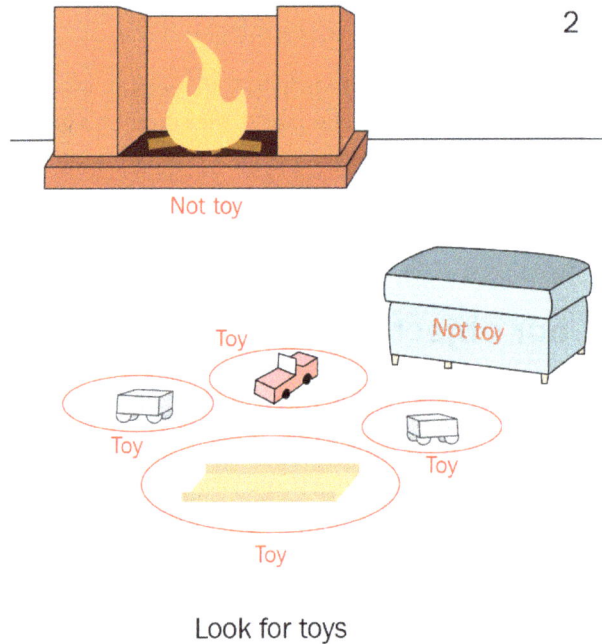

Figure 4.1 – Use case for identifying objects as toys

Our image recognizer must figure out what objects are toys and then locate them in the image. This is illustrated in the preceding sketch; objects marked as toys have circles around them. The image recognizer must recognize not just *what* they are, but also *where* they are.

Image manipulation

So, now that we have an image, what can we do with it? You have probably played with Adobe Photoshop or some other image manipulation program such as GIMP, and you know that there are hundreds of operations, filters, changes, and tricks you can perform on images. For instance, can make an image brighter or darker by adjusting the brightness. We can increase the contrast between the white parts of the image and the dark parts. We can make an image blurry, usually by applying a Gaussian blur filter. We can also make an image sharper (somewhat) by using a filter such as an unsharp mask. You can also use an edge detector filter, such as the Canny filter, to isolate the edges of an image, where color or value changes. We will be using all of these techniques to help the computer identify images:

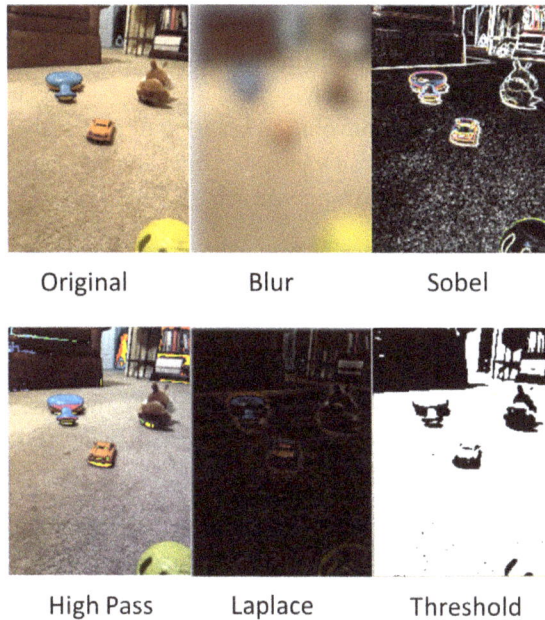

Figure 4.2 – Various convolutions applied to an image

By performing these manipulations, we want the computer to not have the computer software be sensitive to the size of the image, which is called **scale invariance**, the angle at which the photograph was taken, or **angle invariance**, and the lighting available, which is known as **illumination invariance**. This is all very desirable in a computer vision system – we would not want an AI system that only recognizes our toys from the same angle and distance as the original image. Remember, we are going to train our vision system to recognize toys based on a labeled set of training images we take in advance, and the robot will have to recognize objects based on what it learned from the training set. Here, we are going to use features from the image that mostly don't change based on size, angle, distance, or lighting. What sorts of features might these be?

If we look at a common household object, such as a chair, and inspect it from several angles, what about the chair does not change? The easy answer is the edges and corners. The chair has the same number of corners all the time, and we can see a consistent number of them from most angles. It also has a consistent number of edges.

Admittedly, that is a bit of an oversimplification of the approach. We will be training our ANN on a whole host of image features that may or may not be unique to this object and let it decide which work and which do not. We will accomplish this by using a generic approach to image manipulation called **convolution**.

Convolution

Every once in a while, you'll come across some mathematical construction that turns a complex task into just a bunch of adding, subtracting, multiplying, and dividing. Vectors in geometry work like that, and, in image processing, we have the **convolution kernel**. It transpires that most of the common image processing techniques – edge detection, corner detection, blurring, sharpening, enhancing, and so on – can be accomplished with a simple array construct.

It is pretty easy to understand that in an image, the neighbors of a pixel are just as important to what a pixel is as the pixel itself. If you were going to try and find all the edge pixels of a box, you would look for a pixel that has one type of color on one side, and another type on the other. We need a function to find edges by comparing pixels on one side of a pixel to the other.

The convolution kernel is a matrix function that applies weights to the pixel neighbors – or pixels around the one pixel we are analyzing. The function is usually written like this, as a 3x3 matrix:

-1	0	1
-2	0	2
-1	0	1

Table 4.1 – A sample convolution kernel

Sobel edge detection is represented in the Y direction. This detects edges going up and down. Each block represents a pixel. The pixel being processed is in the center. The neighbors of the pixels on each side are the other blocks – top, bottom, left, and right. To compute the convolution, the corresponding weight is applied to the value of each pixel by multiplying the value (intensity) of that pixel, and then adding all of the results. If this image is in color – RGB – then we compute the convolution for each color separately and then combine the results. Here is an example of a convolution being applied to an image:

Figure 4.3 – Result of a Sobel edge detection convolution

The resulting image is the same size as the original. Note that we only get the edge as the result – if the colors are the same on either side of the center pixel, they cancel each other out and we get zero, or black. If they are different, we get 255, or white, as the answer. If we need a more complex result, we may also use a 5x5 convolution, which takes into account the two nearest pixels on each side, instead of just one.

The good news is that you don't have to choose which convolution operation to apply to the input images – we will build a software frontend that will set up all of the convolutions. This *frontend* is just the part of the program that sets up the networks before we start training them. The neural network package we'll be using will determine which convolutions provide the most data and support the training output we want.

"But wait," I hear you say. "What if the pixel is on the edge of the image and we don't have neighbor pixels on one side?" In that case, we have to add padding to the image – which is a border of extra pixels that permits us to consider the edge pixels as well.

In the next section, we'll get into the guts of a neural network.

Artificial neurons

What is a **neuron**? And how do we make a network out of them? If you can remember what you learned in biology, a biological or natural neuron has inputs, or dendrites, that connect it to other neurons or sensor inputs. All the inputs come to a central body and then leave via the axion, or connection, to other neurons via other dendrites. The connection between neurons is called a **synapse**, which is a tiny gap that the signal from the nerve must jump. A neuron takes inputs, processes them, and activates or sends an output after some threshold level is reached. An **artificial neuron** is a software construction that approximates the workings of the neurons in your brain and is a very simplified version of the natural neuron. It has several inputs, a set of weights, a bias, an activation function, and then some outputs to other neurons as a result of the network, as shown in the following figure:

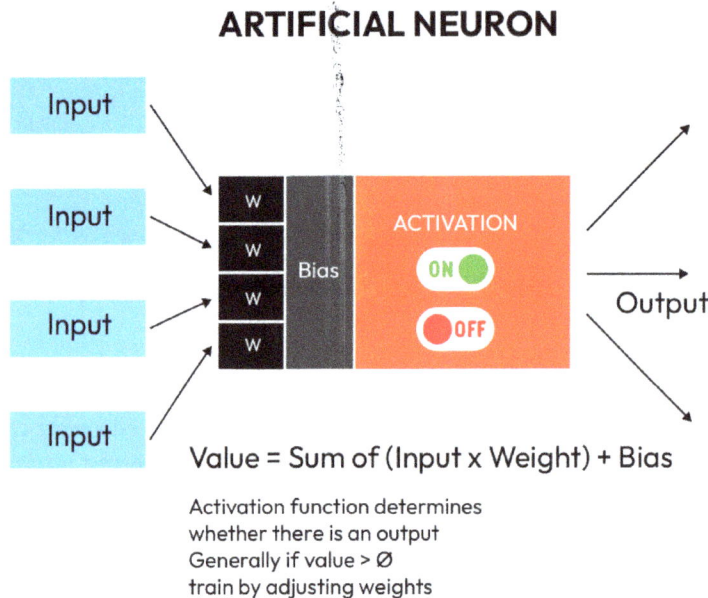

ARTIFICIAL NEURON

Value = Sum of (Input x Weight) + Bias

Activation function determines
whether there is an output
Generally if value > Ø
train by adjusting weights

Figure 4.4 – Diagram of an artificial neuron

Let's describe each component in detail:

- **Input**: This is a number or value that's received from other neurons or as an input to the network. In our image processing example, these are pixels. This number can be a float or an integer – but it must be just one number.

- **Weight**: This is the adjustable value we change to *train* the neuron. Increasing the weight means that the input is more important to our answer, and likewise decreasing the weight means the input is used less. To determine the value of a neuron, we must combine the values of all the inputs. As the neural network is trained, the weights are adjusted on each input, which favors some inputs over others. We multiply the input by the weight and then sum all of the results together.

- **Bias**: This is a number that's added to the sum of the weights. Bias prevents the neuron from getting stuck at zero and improves training. This is usually a small number. Imagine a scenario where all of the inputs to a neuron are zero; in this case, the weights would have no effect. Adding a small bias allows the neuron to still have an output, and the network can use that to affect learning. Without the bias, a neuron with zeros on its inputs can't be trained (changing the weights has no effect) and is *stuck*.

- **Activation function**: This determines the output of the neuron based on the weighted sum of its inputs. The most common types are the **Rectified Linear Unit (ReLU)** – if the value of the neuron is less than zero, the output is zero; otherwise, the output is the input value – and the **sigmoid** (S-shaped) function, which is a log function. The activation function propagates information across the network and introduces non-linearity to the output of the neuron, which allows the neural network to approximate non-linear functions:

COMMON ACTIVATION FUNCTIONS

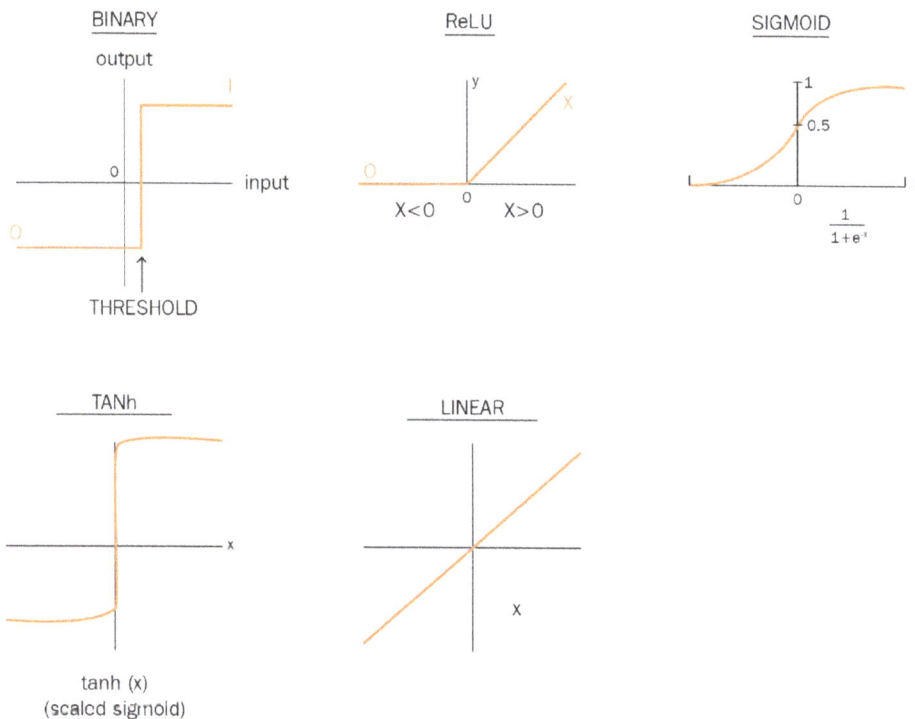

Figure 4.5 – Common activation functions

- **Outputs**: Each layer in the sequential neural network is connected to the next layer. Some layers are fully connected – with each neuron in the first layer connected to each neuron in the second layer. Others are sparsely connected. There is a common process in neural network training called **dropout**, where we randomly remove connections. This forces the network to have multiple paths for each bit of information it learns, which strengthens the network and makes it able to handle more diverse inputs.

- **Max pooling of outputs**: We use a special type of network layer (compared to a fully connected or sparse layer) called max pooling, where groups of neurons corresponding to regions in our image – say a 2x2 block of pixels – go to one neuron in the next level. The max pool neuron only takes the largest value from each of the four input neurons. This has the effect of downsampling the image (making it smaller). This allows the network to associate small features (such as the wheels in a Hot Wheels car) with larger features, such as the hood or windshield, to identify a toy car:

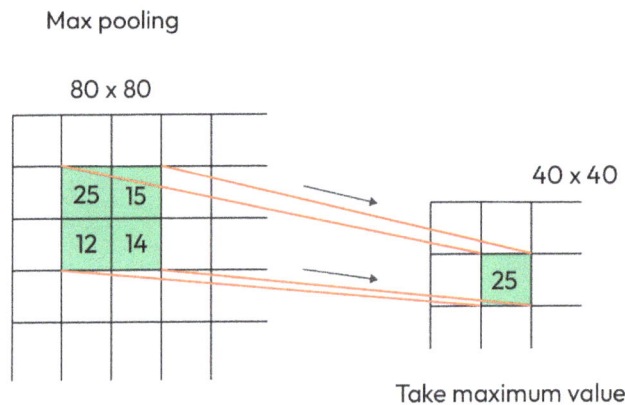

Figure 4.6 – Max pooling operation

Now that you understand what a neural network is composed of, let's explore how to train and test one.

Training a CNN

I want to provide you with an end-to-end look at what we will be doing in the code for the rest of this chapter. Remember that we are building a CNN that examines pixels in a video frame and outputs if one or more pixel areas that resemble toys are in the image, and where they are. The following diagram shows the process that we will go through to train the neural network, step by step:

CONVOLUTIONAL NEURAL NETWORK PROCESS

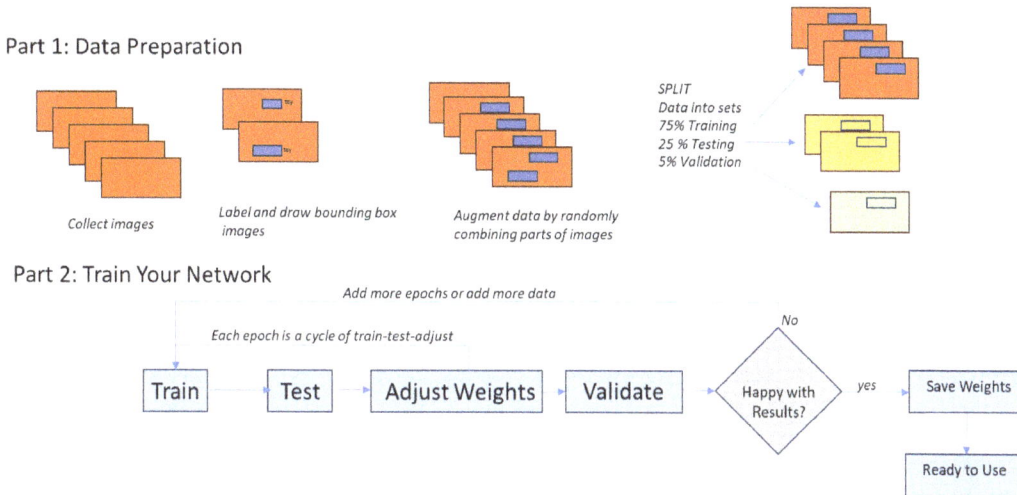

Part 1: Data Preparation

SPLIT
Data into sets
75% Training
25 % Testing
5% Validation

Collect images

Label and draw bounding box
images

Augment data by randomly
combining parts of images

Part 2: Train Your Network

Add more epochs or add more data

Each epoch is a cycle of train-test-adjust

No

Train → Test → Adjust Weights → Validate → Happy with Results? → yes → Save Weights

Ready to Use

Figure 4.7 – CNN process

For this task, I decided to use an already existing neural network rather than build one from scratch. There are a lot of good CNN object detectors available, and honestly, it's hard to improve on an existing model structure. The one I've picked for this book is called **YOLOv8**, where *YOLO* stands for *You Only Look Once*. Let's understand how we can use this model for our task.

Using YOLOv8 – an object recognition model

Before we dive into the details of the YOLOv8 model, let's talk about why I selected it. First of all, the learning process is pretty much the same for any CNN we might use. YOLO is a strong open source object detection model with a lot of development behind it. It's considered state of the art, and it already does what we need – it detects objects and shows us where they are in images by drawing bounding boxes around them. So, it tells us what objects are, and where they are located. As you will see, it is very easy to use and can be extended to detect other classes of objects other than what it was originally trained for. There are a lot of YOLO users out there who can provide a lot of support and a good basis for learning about AI object recognition for robotics.

As I mentioned at the beginning of this chapter, we have two tasks we need to accomplish to reach our goal of picking up toys with a robot. First, we must determine whether the robot can detect a toy with its camera (determine whether there is a toy in the camera image) and then figure out where it is in that image so that we can drive over to it and pick it up. In this chapter, we'll learn how to detect toys, while in *Chapter 7*, we'll discuss how we determine distance and navigate to the toy.

YOLOv8 does both tasks in one pass, hence its name. Other kinds of object recognition models, such as the one I created in the first edition of this book, identified and located objects in images in two steps. First, it found that an object was present, and then it figured out where in the image it was located in a separate pass. This separate pass would use a sliding window approach, taking a segment of the image and using the detection part of the neural network to say yes or no if that segment contained an object it recognized. Then, it would slide the window it was considering across the image and test again. This was repeated until we had a bunch of image segments that contained the detected object. Then, a process called *minmax* would select the smallest box (min) that contained all the visible parts of the object (max).

YOLOv8 takes a different approach by combining two neural networks – one that detects objects it has been taught to recognize and another that is trained to draw bounding boxes based on the center of the object. The direct output of YOLOv8 includes both the detection and the bounding box for the object. YOLOv8 can also *segment* images by pixels, identifying not just a box with the object, but all the pixels that belong to that object. We'll be using a bounding box to help us drive the robot to our toys.

YOLOv8 comes pretrained on a whole series of object classes (about 80), but we can check whether it already works with the toys we want to detect. Let's test YOLOv8's ability to detect our toys. We can install YOLOv8 using this simple command on our PC:

```
pip install ultralytics
```

Now, to test our detection with a picture of toys in the playroom, we will use the smallest (in terms of model size) of the YOLOv8 detection models – the **nano-sized model** (which is called yolov8n. pt). This is the pretrained neural network that Ultralytics provides with YOLOv8:

```
yolo task=detect mode=predict model=yolov8n.pt source="test.
png"
```

As shown in the following figure, the only thing detected by the off-the-shelf YOLOv8 object model is an upside-down matchbook car, which it incorrectly labels as a skateboard:

Figure 4.8 – YOLOv8 output without specific training on our toys

You have to admit, the little toy car does resemble a skateboard from this angle, but this is not the result we want. We need all the toys in the image to be detected, not just one. What can we do about this?

The answer is that we can add new training to the network, get all the advantages of YOLOv8, and have our custom objects detected as well. For this, we can use a process called **transfer learning**.

Here is an overview of how we will train our toy detector, after which we will discuss these steps in greater detail:

1. First, we will prepare a training set of images of the room with toys. This means we must take a lot of pictures of the toys from the viewpoint of the robot, using the same camera the robot will use. We want to take pictures from all the different angles and sides of the toys. I went around the room clockwise, then anti-clockwise, snapping pictures every few inches. I took 48 pictures in this step.

2. Next, we must use a data labeling program such as RoboFlow (`https://roboflow.com`) to annotate the images (you can refer to the relevant documentation for detailed instructions). The program lets us draw boxes around the objects we want to recognize (toys) and label them with a tag – we will use the name `toy`. We are separating the parts of the image that contain toys and telling the neural network what to call this type of object:

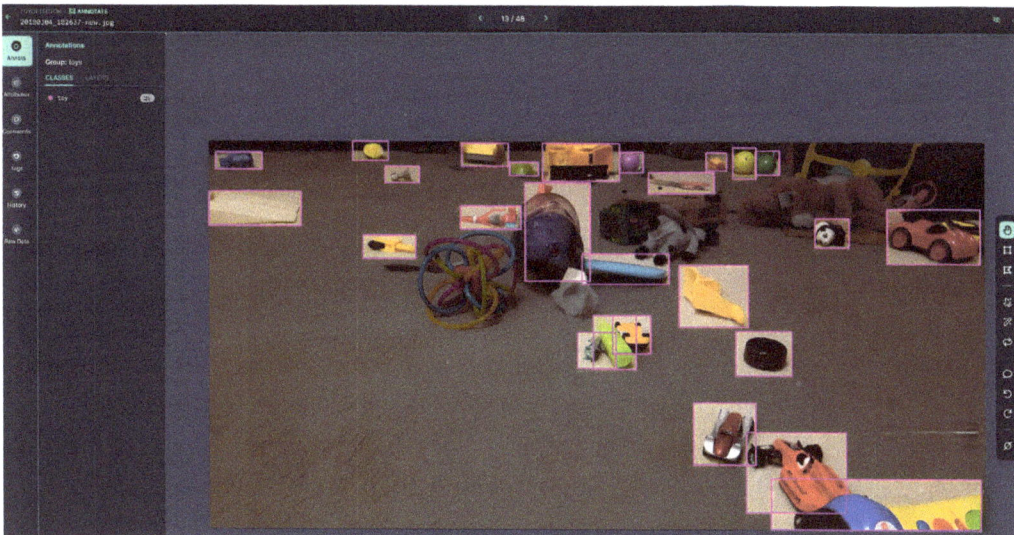

Figure 4.9 – Annotating using RoboFlow, a free data labeling tool

3. Then, we must split the training set into three parts: a set we use to train the network, a set we use to validate the training, and a set we use to test the network. We will create a set of 87% of the images for training, 8% for validation, and 5% for testing. We'll put the training data and test data in separate folders. RoboFlow has a procedure for this under the **Generate** tab, where there is a section labeled **Train/Test Split**.

4. Now, we must take each image and multiply its training value by combining parts of different images in a mosaic. We'll take parts of four random different pictures and combine them. This will increase our training set three-fold, a process called **data augmentation**. This is built into RoboFlow. I started with 36 pictures; after augmentation, I had 99:

Figure 4.10 – Mosaic data augmentation creates more training data from our limited number of pictures

Why are we using this mosaic approach? We still want to have valid bounding boxes. The mosaic process resizes any partial bounding boxes that intersect the edges.

5. Next, we will be building two programs: the training program, which runs on our desktop computer and trains the network, and the working program, which uses the trained network to find toys. The training process may not run on our small computer onboard the robot or may take a long time to run, so we'll use the desktop computer for this.

6. Now, we need to train the network. To achieve this, we must do the following:

    I.   First, we must scale all our images down to reduce the processing time to a reasonable level.

    II. Then, we must initialize the network with uniform random weights.

    III. Next, we must encode a labeled image and input that to the network. The neural network only uses the image data to predict what class of object is in the picture, and what its bounding box should be. Since we pre-labeled the image with the correct answer and used the correct bounding box, we can judge whether the answer is right or wrong. If it is right, we can reinforce the weights on the inputs that contributed to this answer by incrementing them (the training value). If the answer is wrong, we can reduce the weights instead. In neural networks, the error between the desired result and the actual result is called **loss**. Repeat this process for each image.

IV.   Now, we must test the network by running the testing set of images – which are pictures of the same toys, but that were not in the training set. We must analyze what sort of output we get over this set (how many are wrong and how many are right). If this answer is above 90%, we stop. Otherwise, we go back and run all the training images again.

V.   Once we are happy with the results – and we should need between 50 and 100 iterations to get there – we must stop and store the weights that we ended up with in the training network. This is our **trained CNN**.

7.   Our next task is to find the toys. To do this, we must *deploy* the trained network by loading it and using our video images from the live robot to look for toys. We'll get a probability of an image containing a toy from 0% to 100%. We'll scan the input video image in sections and find which sections contain toys. If we are not happy with this network, we can reload this network into the training program and train it some more.

Now, let's cover this in detail, step by step. We have a bit more theory to cover before we start writing the code.

Understanding how to train our toy detector

Our first task is to prepare a training set. We'll put the camera on the robot and drive the robot around using the teleoperation interface (or just by pushing it around by hand), snapping still photos every foot or so. We just need pictures with toys in the image since we will be annotating the toys. We need about 200 pictures – the more, the better. We also need to have a set of pictures in the daytime with natural light, and at night, if your room changes lighting between day and night. This affords us several advantages: we are using the same room and the same camera to find the toys, all under the same lighting conditions.

Now, we need to label the images. We'll load the images into RoboFlow to create a dataset called `toydetector`. Use the **Upload** tab and drag and drop the images or select the folder that contains the images.

The process for us is fairly straightforward. We look at each picture in turn and draw a box around any toy objects. We hit *Enter* or the **Save** button. The annotation dialog box will open. As you might have guessed, we'll be labeling these `toy`. This is going to take a while.

Once we've labeled around 160 toys in our images, we can use the **Generate** button in RoboFlow to create our dataset. We must set up the preprocessing task to resize our images to 640x640 pixels. This makes the best use of our limited computer capacity on the robot. Then, we must augment the dataset to create additional images of our limited set, as mentioned previously. We'll use the mosaic method to augment our dataset while preserving the bounding boxes. To do this, we must use the **Generate** tab in RoboFlow, then click **Add Augmentation Step** to select the type of operation that will affect our images. Then, we must add the mosaic augmentation to create more images out of our training set. Now, we can hit the **Generate** button to create our dataset.

We started with 48 images that I took (back in step 1); after augmentation, we have 114. We'll set our test/train split so that it contains 99 images in the training set, nine images in the validation set, and six images in the test set (87% training, 8% validation, and 5% testing). This makes the best use of our limited dataset.

To download our datasets from RoboFlow, we must install RoboFlow's interface on our computer. It's a Python package:

```
pip install roboflow
```

Then, we must create a short Python program called downloadDataset.py. When you build your dataset, RoboFlow will provide a unique api_key value; this will be the password for your account that authorizes access. This goes into the program as follows, where I put the asterisks:

```
from roboflow import Roboflow
rf = Roboflow(api_key="*****************")
project = rf.workspace("toys").project("toydetector")
dataset = project.version(1).download("yolov8")
```

In the next section, we will retrain the network with this command:

```
yolo task=detect mode=train model=yolov8n.pt data=datasets/
data.yaml epochs=100 imgsz=640
```

Once we've done this, the program will produce a lot of output, as follows:

```
(p310) E:\BOOK\YOLO>yolo task=detect mode = val model=runs\
detect\train3\weights\best.pt data=ToyDetector-1\data.yaml
Ultralytics YOLOv8.0.78 Python-3.10.10 torch-2.0.0 CUDA:0
(NVIDIA GeForce RTX 2070, 8192MiB)
Model summary (fused): 168 layers, 3005843 parameters, 0
gradients, 8.1 GFLOPs
.................
AMP: checks passed
optimizer: SGD(lr=0.01) with parameter groups 57
weight(decay=0.0), 64 weight(decay=0.0005), 63 bias
train: Scanning E:\BOOK\YOLO\datasets\ToyDetector-1\train\
labels.cache… 99 images, 0 backgrounds, 0 corrupt: 100%|███
```

```
val: Scanning E:\BOOK\YOLO\datasets\ToyDetector-1\valid\labels.
cache… 9 images, 0 backgrounds, 0 corrupt: 100%|██████|
Plotting labels to runs\detect\train5\labels.jpg…
Image sizes 640 train, 640 val
Using 8 dataloader workers
Logging results to runs\detect\train5
Starting training for 100 epochs…
```

One of the critical parts of training our model is the **training optimizer**. We will use **stochastic gradient descent (SGD)** for this. SGD is another of those simple concepts with a fancy name. Stochastic just means *random*. What we want to do is tweak the weights of our neurons to give a better answer than we got the first time – this is what we are training, by adjusting the weights. We want to change the weights a small amount – but in which direction? We want to change the weights in the direction that improves the answer – it makes the prediction closer to what we want it to be.

To understand this better, let's do a little thought experiment. We have a neuron that we know is producing the wrong answer and needs adjusting. We'll add a small amount to the weight and see how the answer changes. It gets slightly worse – the number is further away from the correct answer. So, we must subtract a small amount instead – and, as you might think, the answer gets better. We have reduced the amount of error slightly. If we make a graph of the error produced by the neuron, we'll see that we are moving toward an error of zero, or the graph is descending toward some minimum value. Another way of saying this is that the slope of the line is negative – going toward zero. The amount of the slope can be called a **gradient** – just as you would refer to the slope or steepness of a hill as the gradient. We can calculate the partial derivative (in other words, the slope of the error curve near this point), which tells us the slope of the line.

The way we go about adjusting the weights on the network as a whole to minimize the loss between the ground truth and the predicted value is called **backpropagation**. This is because, as you might surmise, we have to start at the end of the network – where we know what the answer is supposed to be – and work our way toward the beginning. We have to calculate how each neuron contributes to the answer we want and adjust it slightly (this is known as the **learning rate**) in the right direction to move toward the correct answer every time. For this, we must go back to the idea of a neuron – we have inputs, weights for each input, a bias, and then an activation function to produce an output. If we know what the output is, we can work backward through the neuron to adjust the weights. Let's take a simple example. We have a neuron with three inputs – $Y1$, $Y2$, and $Y3$. We have three weights – $W1$, $W2$, and $W3$. We'll have the bias, B, and our activation function, D, which is the ReLU rectifier. The values of our inputs are 0.2, 0.7, and 0.02. The weights are 0.3, 0.2, and 0.5. Our bias is 0.3, and the desired output is 1.0. We calculate the sum of the inputs and weights and we get a value of 0.21. After adding our bias, we get 0.51. The ReLU function passes any value greater than zero, so the activated output of this neuron is 0.51. Our desired value is 1.0, which comes from the truth (label) data. So, our error is 0.49. If we add the training rate value to each weight, what happens? Take a look at the following diagram:

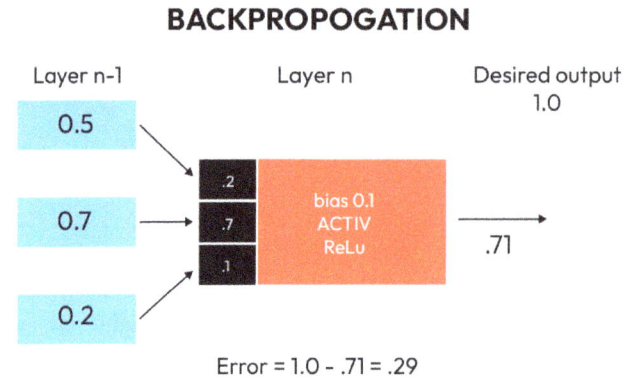

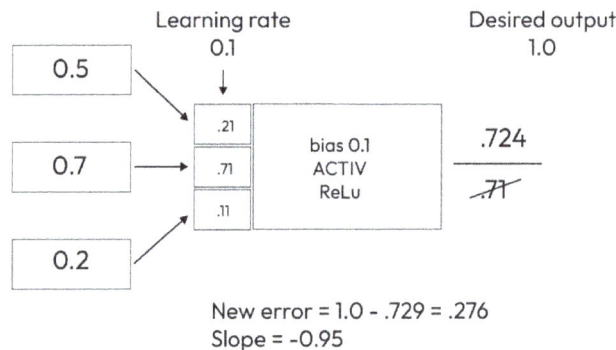

BACKPROPOGATION

Figure 4.11 – How backpropagation works to adjust weights

The output value now goes up to 0.5192. Our error goes down to 0.4808. We are on the right track! The gradient of our error slope is *(.4808-.49) / 1 = -0.97*. The *1* is because we just have one training sample so far. So, where does the stochastic part come from? Our recognition network may have 50 million neurons. We can't be doing all of this math for each one. So, we must take a random sampling of inputs rather than all of them to determine whether our training is positive or negative.

In math terms, the slope of an equation is provided by the derivative of that equation. So, in practice, backpropagation takes the partial derivative of the error between training epochs to determine the slope of the error, and thus determine whether we are training our network correctly. As the slope gets smaller, we reduce our training rate to a smaller number to get closer and closer to the correct answer:

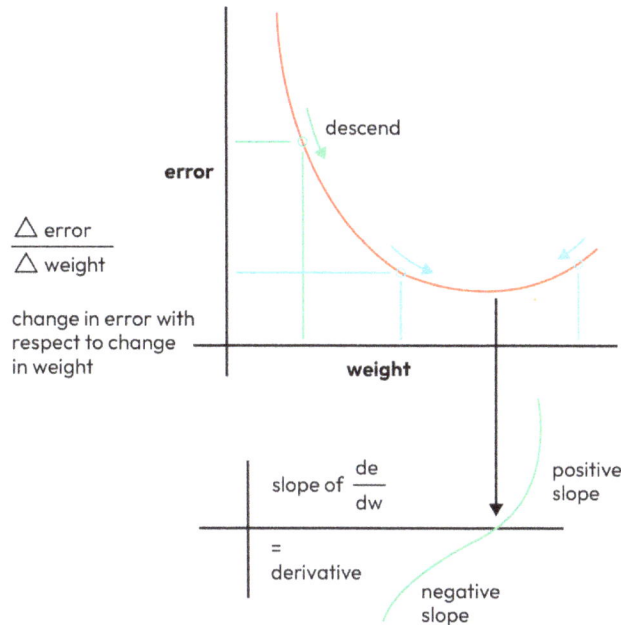

Figure 4.12 – The gradient descent process

Now, we can to our next problem: how do we propagate our weight adjustments up the layers of the neural network? We can determine the error at the output neuron – just the label value minus the output of the network. How do we apply this information to the previous layer? Each neuron's contribution to the error is proportional to its weight. We must divide the error by the weight of each input, and that value is now the applied error of the next neuron up the chain. Then, we can recompute their weights and so on. This is why neural networks take so much compute power:

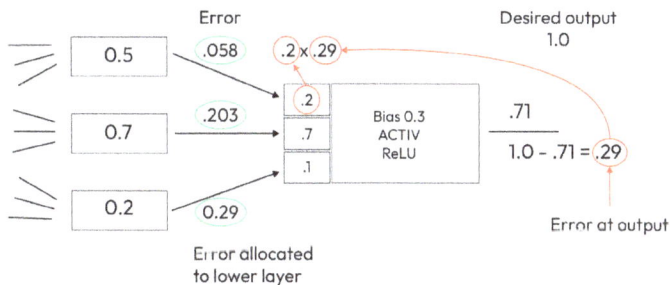

Figure 4.13 – Backpropagation error

We backpropagate the error back up the network from the end back to the beginning. Then, we start all over again with the next cycle.

At this point, we can test our toy detector. Let's see how we can do this.

Building the toy detector

We can use the following command to test how we did:

```
yolo task=detect mode=predict model=last.pt source=toy1.jpg
imgsz=640
```

The program produces the following output. We can find our image with the labeled detections in the `./runs/detect/predict` directory with a number appended depending on how many times we run the detection:

```
Speed: 4.0ms preprocess, 44.7ms inference, 82.6ms postprocess
per image at shape (1, 3, 640, 640)
Results saved to runs\detect\predict4
```

The output of our prediction is shown in the following figure:

Figure 4.14 – The toy detector in action

With this, we have successfully created a toy detector using a neural network. The output of the detector, which we will use in *Chapter 5* to direct the robot and the robot arm to drive to the toy and then pick it up, looks like this:

```
"predictions": [
 {
 "x": 287.5,
 "y": 722.5,
 "width": 207,
 "height": 131,
 "confidence": 0.602,
 "class": "toy"
 },
```

For each detection, the neural network will provide several bits of information. We get the x and y locations of the center of the bounding box, and then the height and width of that box. Then, we get a confidence number of how certain the network is of the decision that this is a detection. Finally, we get the class of the object (what kind of object), which is, of course, a toy.

When we ran the training process for the neural network, if you look in the `training` folder found in `runs/detect/train`, there are a whole series of graphs. What do these graphs tell us?

The first one we need to look at is `F1_curve`. This is the product of precision and recall. **Precision** is the ratio of true positives (correctly classified objects) from all positives. **Recall** is the proportion of positive detections that were identified correctly. So, precision is defined as follows:

$Precision = TP\_TP + FP$

Precision is the true positives divided by true positives and false positives (items that were identified as detections but were not).

Recall is defined slightly differently:

$Recall = TP\_TP + FN$

Here, recall is the true positives divided by true positives plus false negatives. A false negative is a missed detection or an object that was not detected when it was, in fact, present.

To create the F1 curve, we must multiply precision and recall together and plot it against *confidence*. The graph shows the level of confidence in our detections that produces the best result of trading off precision and recall:

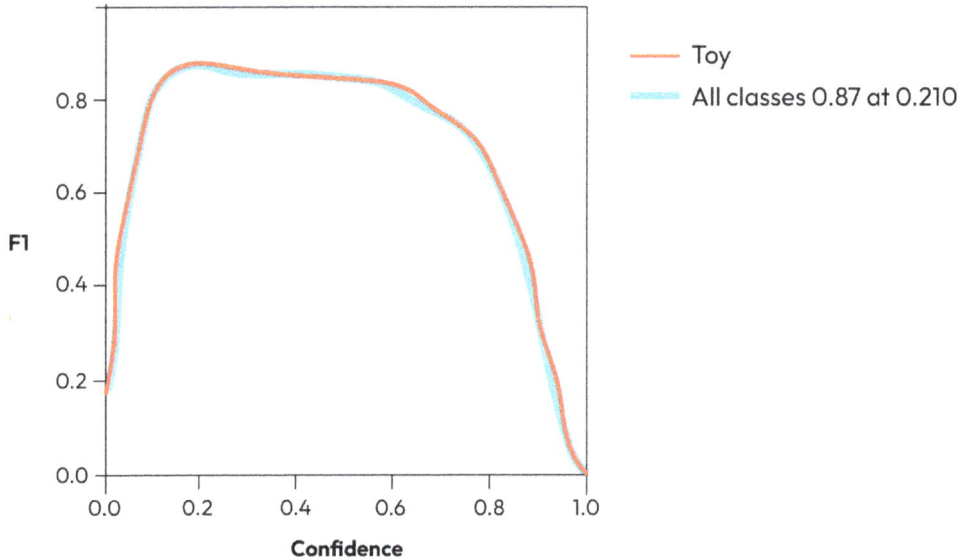

Figure 4.15 – The F1 confidence curve

In this case, a confidence level of 0.21 gives a detection rate of 0.87. This means that we get the best ratio of true detections to false detections. However, this best ratio – 87% – occurs at 0.21 confidence – a rather low number. Detections at this low confidence level are hard to distinguish and can be caused by noise in measurements. It might be more desirable to have our peak at a higher confidence level. I tried several approaches to address this. I ran 200 epochs rather than 100 and moved the peak F1 confidence level to 51%, but the detection level dropped a bit to 85%. Then, I changed the gradient descent technique from SDM to **Adam**, an adaptive gradient descent technique that reduces the learning rate as you get closer to our goal. This can be done using the following code:

```
yolo task=detect mode=train model=yolov8n.pt data=datasets/
data.yaml epochs=100 optimizer='adamW' imgsz=640
```

This produced a more satisfactory result of 88% true detections at 49% confidence, which I think will do a better job for our toy detector. In reviewing my detections, there were a few false positives (furniture and other objects being detected as toys), so I think that this version will be our toy detector neural network. Although I used a fairly small dataset, it would not hurt to have more pictures to work with from different angles. Before wrapping this chapter up, let's briefly summarize what we've learned so far.

Summary

In this chapter, we dove head-first into the world of ANNs. An ANN can be thought of as a stepwise non-linear approximation function that slowly adjusts itself to fit a curve that matches the desired input to the desired output. The learning process consists of several steps, including preparing data, labeling data, creating the network, initializing the weights, creating the forward pass that provides the output, and calculating the loss (also called the error). We created a special type of ANN, a CNN, to examine images. The network was trained using images with toys, to which we added bounding boxes to tell the network what part of the image was a toy. We trained the network to get an accuracy better than 87% in classifying images with toys in them. Finally, we tested the network to verify its output and tuned our results using the Adam adaptive descent algorithm.

In the next chapter, we will look at machine learning for the robot arm in terms of reinforcement learning and genetic algorithms.

Questions

1. We went through a lot in this chapter. You can use the framework provided to investigate the properties of neural networks. Try several activation functions, or different settings for convolutions, to see what changes in the training process.

2. Draw a diagram of an artificial neuron and label the parts. Look up a natural, human biological neuron and compare them.

3. Which features of a real neuron and an artificial neuron are the same? Which ones are different?

4. What effect does the learning rate have on gradient descent? What if the learning rate is too large? Too small?

5. What relationship does the first layer of a neural network have with the input?

6. What relationship does the last layer of a neural network have with the output?

7. Look up three kinds of loss functions and describe how they work. Include mean square loss and the two kinds of cross-entropy loss.

8. What would you change if your network was trained and reached 40% accuracy of the classification and got stuck, or was unable to learn anything further?

Further reading

For more information on the topics that were covered in this chapter, please refer to the following resources:

- *Python Deep Learning Cookbook*, by Indra den Bakker, Packt Publishing, 2017
- *Artificial Intelligence with Python*, by Prateek Joshi, Packt Publishing, 2017

- *Python Deep Learning*, by Valentino Zocca, Gianmario Spacagna, Daniel Slater, and Peter Roelants, Packt Publishing, 2017

- *PyImageSearch Blog*, by Adrian Rosebrock, available at `pyimagesearch.com`, 2018

5

Picking Up and Putting Away Toys using Reinforcement Learning and Genetic Algorithms

This chapter is where the robots start to get challenging – and fun. What we want to do now is have the robot's manipulator arm start picking up objects. Not only that, but instead of preprogramming arm moves and grasping actions, we want the robot to be able to learn how to pick up objects, and how to move its arm without hitting itself.

How would you teach a child to pick up toys in their room? Would you offer a reward for completing the task, such as *"If you pick up your toys, you will get a treat?"* Or would you offer a threat of punishment, such as *"If you don't pick up your toys, you can't play games on your tablet."* This concept, offering positive feedback for good behavior and negative feedback for undesirable actions, is called **reinforcement learning**. That is one of the ways we will train our robot in this chapter.

If this sounds something like a game, where you get positive points for reaching a goal and lose points for missing a goal, then you are right. We have some concept of winning that we are trying to achieve, and we create some sort of point system to reinforce – that is to say, reward – behavior when the robot does what we want it to.

In this chapter, we will cover the following topics:

- Designing the software
- Setting up the solution
- Introducing Q-learning for grasping objects
- Introducing **genetic algorithms** (**GAs**) for path planning
- Alternative robot arm ML approaches

Technical requirements

The exercise in this chapter does not require any new software or tools that we haven't already seen in previous chapters. We will start by using Python and ROS 2. You will need an IDE for Python (IDLE or Visual Studio Code) to edit the source code.

Since this chapter is all about moving the robot arm, you will need a robot arm to execute the code. The one I used is the **LewanSoul Robot xArm**, which I purchased from Amazon.com. This arm uses digital servos, which makes the programming much easier, and provides us with position feedback, so we know what position the arm is in. The arm I purchased can be found at `http://tinyurl.com/xarmRobotBook` at the time of publication.

> **Note**
> If you don't want to buy a robot arm (or can't), you can run this code against a simulation of a robot arm using ROS 2 and **Gazebo**, a simulation engine. You can find instructions at `https://community.arm.com/arm-research/b/articles/posts/do-you-want-to-build-a-robot`.

You'll find the code for this chapter in the GitHub repository for this book, at `https://github.com/PacktPublishing/Artificial-Intelligence-for-Robotics-2e`.

Task analysis

Our tasks for this chapter are pretty straightforward. We will use a robot arm to pick up the toys we identified in the previous chapter. This can be divided into the following tasks:

- First, we build an interface to control the robot arm. We are using **ROS 2** to connect the various parts of the robot together, so this interface is how the rest of the system sends commands and receives data from the arm. Then we get into teaching the arm to perform its function, which is picking up toys. The first level of capability is picking up or grasping toys. Each toy is slightly different, and the same strategy won't work every time. Also, the toy might be in different orientations, so we have to adapt to how the toy is presented to the robot's end effector (a fancy name for its hand). So rather than write a lot of custom code that may or may not work all the time, we want to create a structure so that the robot can learn for itself.

- The next problem we face is to have the arm move. It's not just that the arm has positions, but it also has to have a path from a start point to an end point. The arm is not a monolithic part – it's composed of six different motors (as shown in *Figure 5.3*) that each do something different. Two of the motors – the grip and the wrist – don't move the arm at all; they only affect the hand. So our arm path is controlled by four motors. The other big problem is that the arm can collide with the body of the robot if we are not careful, so our path planning for the arm has to avoid collisions.

We will use a completely different technique for learning arm paths. A **GA** is a technique for machine learning that uses an analog of evolution to *evolve* complex behaviors out of simple movements.

Now let's talk a bit first about what we have to work with. We have a **mobile base** with a six-degrees-of-freedom arm attached to it. We are fortunate that Albert's robot arm is constructed out of digital servo motors. We can tell where all of the parts of the arm are because we can read the position of the servo motors. We also know what we commanded the servo to do. Servo position is reported in units from 0 to 1,024 that represent from 0 degrees to 360 degrees of rotation. If we command a servo to move to position 600, and (after a short time interval to permit the motor to move) we see that the servo position is 421, then something is preventing the motor from reaching the goal we set for it. This information will be very valuable for training the robot arm.

We can use **forward kinematics**, which means summing up all the angles and levers of the arm to deduce where the hand is located (I'll provide the code for that later in the chapter). We can use this hand location as our desired state – our **reward criteria**. We will give the robot points, or rewards, based on how close the hand is to the desired position and orientation we want. We want the robot to figure out what it takes to get to that position. We need to give the robot a way to test out different theories or actions that will result in the arm moving.

We will begin by just working with the robot hand, or to use the fancy robot term, the **end effector**.

The following diagram shows how we are trying to align our robot arm to pick up a toy by rotating the wrist:

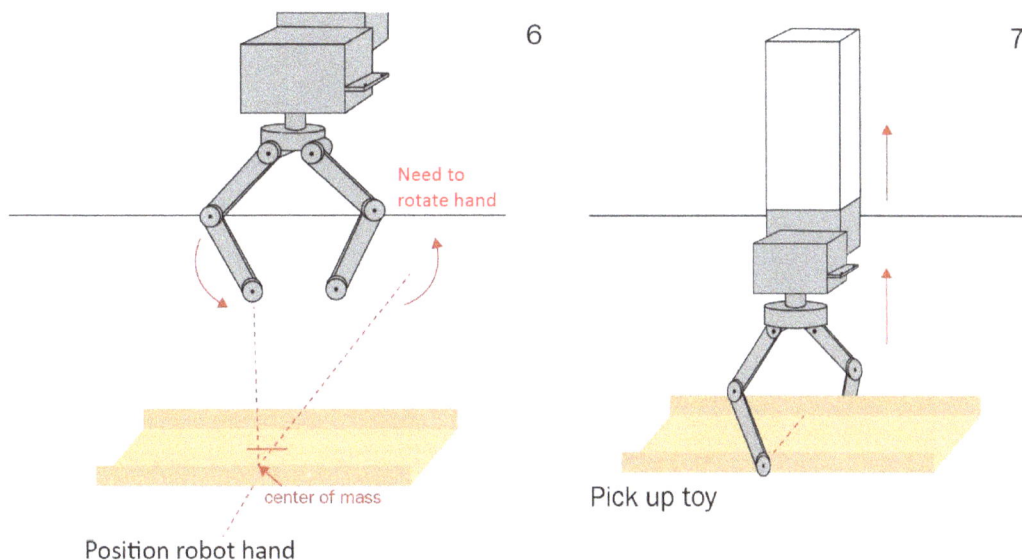

Figure 5.1 – Storyboard for picking up a toy

For grasping, we have three actions to work with. We position the arm to pick up the toy, we adjust the angle of the hand by rotating the hand with the wrist servo, and we close the hand in order to grasp the object. If the hand closes completely, then we missed the toy and the hand is empty. If the toy is keeping the gripper from closing because we picked it up, then we have success and have grabbed the toy. We'll be using this process to teach the robot to use different hand positions to pick up toys based on their shape.

Designing the software

The first steps in designing the robot arm control software are to establish a coordinate frame (how we measure movement), after which we set up our solution space by creating states (arm positions) and actions (movements that change positions). The following diagram shows the coordinate frame for the robot arm:

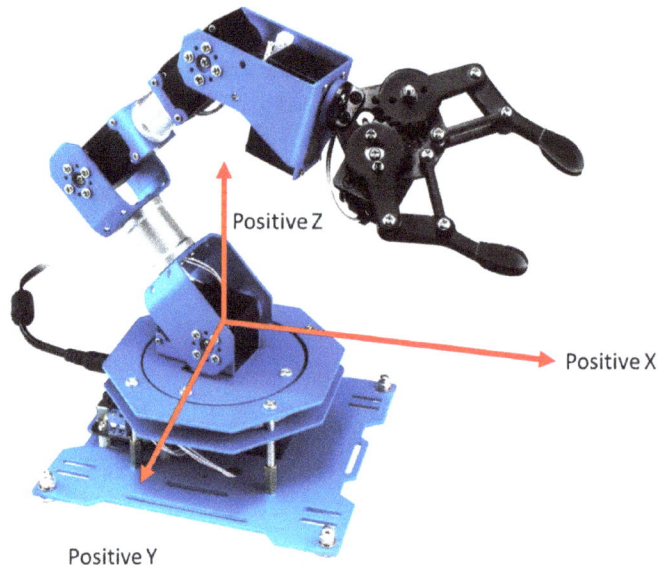

Figure 5.2 – Robot arm coordinate frame

Let's define the coordinate frame of the robot – our reference that we use to measure movement – as shown in the preceding diagram. The X direction is toward the front of the robot, so movement forward and backward is along the X-axis. Horizontal movement (left or right) is along the Y-axis. Vertical movement (up and down) is in the Z direction. We place the zero point – the origin of our coordinates – down the center of the robot arm with zero Z (Z=0) on the floor. So, if I say the robot hand is moving positively in X, then it is moving away from the front of the robot. If the hand (the end of the arm) is moving in Y, then it is moving left or right.

Now we must have a set of names that we will call the servo motors in the arm. We'll do a bit of anthropomorphic naming, and give the arm parts anatomical titles. The motors are numbered in the control system and the servos on my robot arm are labeled:

- *Motor 1* opens and closes the gripper. We may also call the gripper the hand.

- *Motor 2* is wrist rotate, which rotates the hand.

- *Motor 3* is the wrist pitch (up and down) direction.

- We'll call *Motor 4* the elbow. The elbow flexes the arm in the middle, just as you expect.

- *Motor 5* is the shoulder pitch servo, which moves the arm up and down, rotating around the Y-axis when the arm is pointing straight ahead.

- *Motor 6* is at the base of the arm, so we'll call it the shoulder yaw (right or left) servo. It rotates the entire arm about the Z-axis. I've decided not to move this axis, since the entire base of the robot can rotate due to the omni wheels. We'll just move the arm up and down to simplify the problem. The navigation system we develop in *Chapter 8* will point the arm in the correct direction.

We will start by defining an interface to the robot arm that the rest of the robot control system can use:

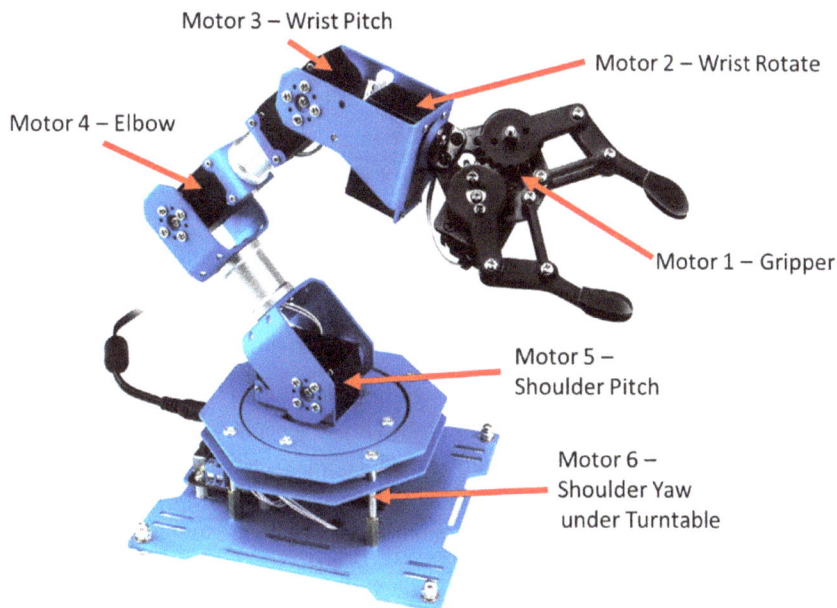

Figure 5.3 – Robot arm motor nomenclature

Here, pitch refers to up/down motion while yaw refers to right/left motion.

We'll use two terms that are common in the robot world to describe how we calculate where the arm is based on the data we have:

- **Forward Kinematics (FK)** is the process of starting at the base of the robot arm and working out toward the gripper, calculating the position and orientation of each joint in turn. We take the position of the joint and the angle it is at, and add the length of the arm between that joint and the next joint. The process of doing this calculation, which produces an X-Y-Z position and a pitch-roll-yaw orientation of the end of the robot's fingers, is called forward kinematics because we calculate forward from the base and out to the arm.

- **Inverse Kinematics (IK)** takes a different approach. We know the position and orientation of either where the hand is, or where we want it to be. Then we calculate backward up the arm to determine what joint angles would produce that hand position. IK is a bit trickier because there may be more than one solution (combination of joint positions) that may produce a given hand result. Try this with your own arm. Grasp a doorknob. Now move your arm while keeping your hand on the doorknob. There are multiple combinations of your joints that result in your hand being in the same position and orientation. We won't be using IK here in this book, but I wanted you to be familiar with the term, which is often used in robot arms to drive the position of robot end effectors (grippers or hands).

For a more in-depth explanation of these concepts, you can refer to `https://control.com/technical-articles/robot-manipulation-control-with-inverse-and-forward-kinematics/`.

Next, let's discuss how we can put the arm in motion.

Setting up the solution

We will call the act of setting the motors to a different position an **action**, and we will call the position of the robot arm and hand the **state**. An action applied to a state results in the arm being in a new state.

We are going to have the robot associate states (a beginning position of the hand) and an action (the motor commands used when at that state) with the probability of generating either a positive or negative **outcome** – we will be training the robot to figure out which sets of actions result in maximizing the **reward**. What's a reward? It's just an arbitrary value that we use to define whether the learning the robot accomplished was positive – something we wanted – or negative – something we did not want. If the action resulted in positive learning, then we increment the reward, and if it does not, then we decrement the reward. The robot will use an algorithm to both try and maximize the reward, and to incrementally learn a task.

Let's understand this process better by exploring the role played by machine learning.

Machine learning for robot arms

Since incremental learning was also part of neural networks, we will use some of the same tools we used before in our neural network to propagate a reward to each step in a chain of movements that result in the hand moving to some location. In reinforcement learning, this is called **discounting the reward** – distributing portions of rewards to the step in a multi-step process. Likewise, the combination of a state and an action is called a **policy** – because we are telling the robot, "when you are in this position, and want to go to that position, do this action." Let's understand this concept better by looking more closely at our process for learning with the robot arm:

1. We set our goal position of the robot hand, which is the position of the robot hand in X and Z coordinates in millimeters from the rotational center of the arm.

2. The robot will try a series of movements to try and get close to that goal. We will not be giving the robot the motor positions it needs to get to that goal – the robot must learn. The initial movements will be totally randomly generated. We will restrict the delta movement (analogous to the learning rate from the previous chapter) to some small size so we don't get wild flailing of the arm.

3. At each incremental movement, we will score the movement based on whether or not the arm moved closer to the goal position.

4. The robot will remember these movements by associating the beginning state and the action (movement) with the reward score.

5. Later, we will train a neural network to generate probabilities of positive outcomes based on the inputs of starting state and movement action. This will allow the arm to learn which sequences of movement achieve positive results. Then we will be able to predict which movement will result in the arm moving correctly based on the starting position.

6. You can also surmise that we must add a reward for accomplishing the task quickly – we want the results to be efficient, and so we will add rewards for taking the shortest time to complete the task – or, you could say, we subtract a reward for each step needed to get to the goal so that the process with the fewest steps gets the most reward.

7. We calculate rewards using the **Q-function**, as follows:

$$Q = Q(s,a) + (reward(s,a) \ + g \ * max(Q(s',a')))$$

    where Q represents the reward the robot will get (or expects to get) from a particular action. $Q(s,a)$ is the final reward that we expect overall for an action given the starting state. $reward(s,a)$ is the reward for that action (the small, incremental step we take now). g is a discount function that rewards getting to the goal quicker, that is, with a fewer number of steps (the more steps you have, the more g discounts (removes the reward)), and $max(Q(s',a'))$ selects the action that results in the largest reward out of the set of actions available at that state. In the equation, s and a represent the current state and action, and s' and a' represent the subsequent state and action, respectively. This is my version of Bellman's equation for decision-making, with some adaptations for this particular problem. I added a discount for longer solutions (with more steps, thus taking longer to execute) to reward quicker arm movement (fewer steps), and left out the learning rate (alpha) as we are taking whole steps for each state (we don't have intermediate states to learn).

Next, let's understand how we can teach the robot arm how to learn movement.

How do we pick actions?

What actions can the robot arm perform? As shown in *Figure 5.3*, we have six motors, and we have three options for each motor:

- We can do nothing – that is, not move at all
- We can move counterclockwise, which will make our motor angle smaller
- We can move clockwise, which makes our motor angle larger

> **Note**
>
> Most servo motors treat positive position changes as clockwise rotation. Thus, if we command the rotation to change from 200 to 250 degrees, the motor will turn clockwise 50 degrees.

Our action space for each motion of the robot arm is to move each motor either clockwise, counterclockwise, or not at all. This gives us 729 combinations with 6 motors (3^6 possible actions). That is quite a lot. The software interface we are going to build refers to the robot arm motors by number with *1* being the hand and *6* being the shoulder rotation motor.

Let's reduce this number and just consider the motions of three of the motors – the **shoulder pitch** (*Motor 5*), **elbow pitch** (*Motor 4*), and **wrist pitch** (*Motor 3*). We'll lock the shoulder yaw for now and just move the arm straight ahead, and since rotating the hand and moving the fingers does not change the position or orientation of the hand, we can ignore those, too. Now we have 27 combinations or policies to consider (3^3). We will note this with an **action matrix**. Each action will have three values. An action that reduces the angle of *Motor 5* holds *Motor 4* in place and increases *Motor 3*, noted as [-1, 0, 1]. We will use a value of just +/-1 or 0 in the action matrix to step the motors in small increments. The x-y coordinates of the hand can be computed from the sums of the angles of each joint times the length of the arm.

Here is a Python function to compute the position of the robot hand, given that each arm segment is 10 cm long. You can substitute the length of your robot arm segments. This function turns the motor angles representing the hand position from degrees into x-y coordinates in centimeters:

```
def forward_kinematics(theta1, theta2, theta3, segment_length):
  # Convert degrees to radians
    theta1_rad = math.radians(theta1)
```

```
    theta2_rad = math.radians(theta2)
    theta3_rad = math.radians(theta3)
    # Calculate positions of each joint
    x1 = segment_length * math.cos(theta1_rad)
    y1 = segment_length * math.sin(theta1_rad)
    x2 = x1 + segment_length * math.cos(theta1_rad + theta2_rad)
    y2 = y1 + segment_length * math.sin(theta1_rad + theta2_rad)
    x3 = x2 + segment_length * math.cos(theta1_rad + theta2_rad +
theta3_rad)
    y3 = y2 + segment_length * math.sin(theta1_rad + theta2_rad +
theta3_rad)
return x3, y3
```

The actions of the arm (possible movements) make up the action space of our robot arm, which is the set of all possible actions. What we will be doing in this chapter is investigating various ways of picking which action to perform and when in order to accomplish our tasks, and using machine learning to do it.

Another way of looking at this process is that we are generating a **decision tree**. You are probably familiar with the concept. We have a bit of a unique application when applying this to a robot arm, because our arm is a series of joints connected together, and moving one moves all of the other joints farther out on the arm. When we move Motor 5, Motors 4 and 3 move position in space, and their angles and distances to the ground and to our goal change. Each possible motor move adds 27 new branches to our decision tree, and can generate 27 new arm positions. All we have to do is pick which one to keep.

The rest of this chapter will deal with just how we go about selecting our motions. It's time to start writing some code now. The first order of business is to create an interface to the robot arm that the rest of the robot can use.

Creating the interface to the arm

As previously noted, we are using ROS 2 as our interface service, which creates a **Modular Open System Architecture** (**MOSA**). This turns our components into *plug-and-play* devices that can be added, removed, or modified, much like the apps on a smartphone. The secret to making that happen is to create a useful, generic interface, which we will do now.

> **Note**
> I'm creating my own interface to ROS 2 that is just for this book. We won't be using any other ROS packages with this arm – just what we create, so I wanted the very minimum interface to get the job done.

We'll be creating this interface in Python. Follow these steps:

1. First, create a **package** for the robot arm in ROS 2. A package is a portable organization unit for functionality in ROS 2. Since we have multiple programs and multiple functions for the robot arm, we can bundle them together:

```
cd ~/ros2_ws/src
ros2 pkg create -build-type ament-cmake ros_xarm
colcon build
```

These commands will create a directory structure in our `src` directory where we will store all of the parts we need.

2. We need to install the drivers for xArm so we can use them in Python:

```
pip install xarm
```

3. Now we go to our new source directory:

```
cd ~/ros2_ws/src/ros_xarm/src
```

This is where we will put our code. Let's call the interface to the robot arm `xarm_mgr.py`, which is short for `xarm manager`.

4. Open the editor and let's start coding. First, we need some imports:

```
import rclpy
import xarm
import time
from rlcpy.node import Node
from std_msgs.msg import String, Int32MultiArray, Int32
```

`rclpy` is the ROS 2 Python interface. `xarm` is the interface to the robot arm, while `time` of course is a time module that we will use to set timers. Finally, we use some standard ROS message formats with which to communicate.

5. Next, we are going to create some predefined named positions of the arm as shortcuts. This is a simple way to put the arm where we need it. I've defined five arm preset positions we can call:

Neutral Carry – Normal Arm
Position

Pick Up

Drop Off at
Toy Box

High Carry

Figure 5.4 – Robot arm positions

Let's describe these positions in some detail:

- *High Carry* is the position we want the arm to be at when we are carrying an object such as a toy. The arm is over the robot, and the hand is elevated. This helps to keep the toy from falling out of the hand.

- *Neutral Carry* is the standard position when the robot is driving so that the arm is not in front of the camera.

- *Pick Up* is a combination of *Grasp* and *Grasp Close* (which are not shown individually in the figure). The former is an arm position that puts the hand on the ground so we can pick up an object. The arm is as far out in front of the robot as it will go and touches the ground. The latter just closes the end effector to grab a toy.

- *Drop Off* is the arm position high above the robot to put a toy in the toy box, which is quite tall.

- *Align* (not shown) is a utility mode to check the alignment of the arm. All the servos are set to their middle position and the arm should point at the ceiling in a straight line. If it does not, you need to adjust the arm using the utility that came with it.

Let's see how we can set up the ROS interface. The numbers are the servo motor positions (angles) in units from 0 (fully counterclockwise) to 1000 (fully clockwise). The 9999 code means to not change the servo in that position so we can create commands that don't change the positions of parts of the arm, such as the gripper:

```
HighCarry=[9999,500,195,858,618,9999]
MidCarry=[9999, 500, 500, 807, 443, 9999]
Grasp = [100,500,151,553,117,9999]
GraspClose=[700,9999,9999,9999,9999,9999]
Align=[500,500,500,500,500,500]
```

6. Now we can start defining our robot arm control class. We'll start with the class definition and the initialization function:

```
class xarmControl(Node):
    def __init__(self):
        super().__init__('xarm_manager') # node name
        self.publisher = self.create_publisher(Int32MultiArray,
'xarm_pos', 10)
        self.armAngPub = self.create_publisher(Int32MultiArray,
'xarm_angle', 10)
```

There is quite a bit going on here to set up our ROS interface for the robot arm:

- First of all, we call up the object class structure (super) to initialize our ROS 2 node with the name xarm_manager.

- Then we create a publisher for the arm position information, helpfully called xarm_pos. Here, POS stands for position. This publishes the arm position in servo units, which go from 0 (fully counterclockwise) to 1000 (fully clockwise). We also publish the arm angles in degrees, in case we need that information, as xarm_angle. The center of the servo travel is 0 degrees (500 in servo units). Counterclockwise positions are negative angles while clockwise positions are positive angles. I just used integer degrees (no decimal points) since we don't need that level of precision for the arm. Our High Carry position in servo units is [666,501,195,867,617,500], and in servo angles is [41,0,-76,91,29,0]. We publish our outputs and subscribe to our inputs.

7. Our inputs, or subscriptions, provide the outside interface to the arm. I thought through how the arm might be used, and came up with the interface I wanted to see. In our case, we have a very simple arm, and need only a few commands. First of all, we have a string command called RobotCmd, which lets us create commands to control the mode or state of the robot. This will be used for a lot of commands for the robot, and not just for the arm. I've created several arm mode commands that we'll cover in a few paragraphs. The usefulness of RobotCmd is we can send any string on this input and process it on the receiving end. It's a very flexible and useful interface. Note that for each subscriber, we create a function call to a callback routine.

When the data is published on the interface, the callback routine is called in our program (`xarm_mgr.py`) automatically:

```
self.cmdSubscribe = self.create_subscription(String, 'RobotCmd',
self.cmdCallback,10)
```

8. The next part of the interface allows us to move the base of the arm in yaw, and operate the hand and wrist independently. In this chapter, we are starting with training just the gripper, so it helps to have an independent interface to rotate, open, and close the gripper. Operating the hand does not change the coordinate position of the gripper, so this can be separated. Likewise, we move the hand in yaw – right and left – to line up with toys to be grasped. We are going to start with this function locked off, and we'll add the yaw function later. This is controlled by the computer vision system that we designed in the previous chapter, so it needs a separate interface. We have the `xarmWrist` command to rotate the wrist, `xarmEffector` to open and close the gripper fingers, and `xarmBase` to move the base of the arm right or left:

```
        self.wristSubscribe = self.create_subscription(Int32,
'xarmWrist', self.wristCallback,10)
        self.effSubscribe = self.create_subscription(Int32,
'xarmEffector', self.effCallback,10)
        self.baseSubscribe = self.create_subscription(Int32,
'xarmBase', self.baseCallback,10)
```

9. The last command interface lets us move the arm to any position we specify. Normally, we command the arm to move using an array of numbers, like this: `[100,500,151,553,117,500]`. I've added a *secret feature* to this command. Since we may want to move the arm without either changing the yaw angle (which comes from the vision system) or the hand position (which may or may not be holding a toy), we can send commands that move the arm but don't affect some of the servos, such as the hand. I used the value `9999` as the *don't-move-this-servo* value. So if the arm position command reads `[9999, 9999, 500, 807, 443, 9999]` then the yaw position (*Motor 6*) and the hand position (*Motors 0 and 1*) don't change:

```
        self.baseSubscribe = self.create_subscription(
Int32MultiArray, 'newArmPos', self.moveArmCallback,10)
```

10. Now that we have all of our publish and subscribe interfaces defined, we can open the USB interface to the robot arm and see whether it is responding. If not, we'll throw an error message:

```
        timer_period = 1.0 # seconds
        self.timer = self.create_timer(timer_period, self.timer_
callback)
        self.i = 0 # counter
        try:
            self.arm = xarm.Controller('USB')
            print("ARM OPEN")
        except:
```

```
                    self.get_logger().error("xarm_manager init NO ARM
        DETECTED")
                    self.arm = None
                    print("ERROR init: NO ARM DETECTED")
                return
```

Note

Here is the quick cheat guide to the servos in the xarmPos command array:

```
[grip open/close, wrist rotate, wrist pitch, elbow pitch,
shoulder pitch, shoulder yaw]
```

---

11. Our next function in the source code is to set up a telemetry timer. We want to periodically publish the arm's position for the rest of the robot to use. We'll create a timer callback that executes periodically at a rate we specify. Let's start with once a second. This is an informational value and we are not using it for control – the servo controller takes care of that. This is the code we need:

```
        timer_period = 1.0 # seconds
        self.timer = self.create_timer(timer_period, self.timer_
    callback)
        self.i = 0 # counter
```

The timer_period is the interval between interrupts. The self.timer class variable is a function pointer to the timer function, and we point it at another function, self.timer_callback, which we'll define in the next code block. Every second, the interrupt will go off and call the timer_callback routine.

12. Our next bit of code is part of the hardware interface. Since we are initializing the arm controller, we need to open the hardware connection to the arm, which is a USB port using the **human interface device (HID)** protocol:

```
        try:
            self.arm = xarm.Controller('USB')
            print("ARM OPEN")
        except:
            self.get_logger().error("xarm_manager init NO ARM
        DETECTED")
            self.arm = None
            print("ERROR init: NO ARM DETECTED")
        return
```

We first create a `try` block so we can handle any exceptions. The robot arm may not be powered on, or it may not be connected, so we have to be prepared to handle this. We create an arm object (`self.arm`) that will be our interface to the hardware. If the arm opens successfully, then we return. If not, we run through the `except` routine:

- First, we log that we did not find the arm in the ROS error log. The ROS logging function is very versatile and provides a handy place to store information that you need for debugging.

- Then we set the arm to a null object (`None`) so that we don't throw unnecessary errors later in the program, and we can test to see whether the arm is connected.

13. The next block of code is our timer callback that publishes telemetry about the arm. Remember that we defined two output messages, the arm position and arm angle. We can service them both here:

```
def timer_callback(self):
    msg = Int32MultiArray()
    # call arm and get positions
    armPos=[]
    for i in range(1,7):
        armPos.append(self.arm.getPosition(i))
    msg.data = armPos
    self.publisher.publish(msg)
    # get arm positions in degrees
    armPos=[]
    for i in range(1,7):
        armPos.append(int(self.arm.getPosition(i, True)))
    msg.data = armPos
    #print(armPos)
    self.armAngPub.publish(msg)
```

We are using the `Int32MultiArray` datatype so that we can publish the arm position data as an array of integers. We collect the data from the arm by calling `self.arm.getPosition(servoNumber)`. We append the output to our array, and when we are done, call the ROS publish routine (`self.<topic name>.publish(msg)`). We do the same thing for the arm angle, which we can get by calling `arm.getPosition(servoNumber, True)` to return an angle instead of servo units.

14. Now we can handle receiving commands from other programs. Next, we are going to be creating a control panel for the robot that can send commands and set modes for the robot:

```
def cmdCallback(self, msg):
    self.get_logger().info("xarm rec cmd %s" % msg.data)
    robotCmd = msg.data
    if robotCmd=="ARM HIGH_CARRY":
        self.setArm(HighCarry)
    if robotCmd=="ARM MID_CARRY":
        self.setArm(MidCarry)
    if robotCmd=="ARM GRASP_POS":
        self.setArm(Grasp)
    if robotCmd=="ARM GRASP_CLOSE":
        self.setArm(GraspClose)
    if robotCmd=="ARM ALIGN":
        self.setArm(Align)
```

This section is pretty straightforward. We receive a string message containing a command, and we parse the message to see whether it is something this program recognizes. If so, we process the message and perform the appropriate command. If we get ARM MID_CARRY, which is a command to position the arm to the middle position, then we send a setArm command using the MidCarry global variable, which has the servo positions for all six motors.

15. Next, we write the code for the robot to receive and execute the wrist servo command, which rotates the gripper. This command goes to *Motor 2*:

```
def wristCallback(self, msg):
    try:
        newArmPos = int(msg.data)
    except ValueError:
        self.get_logger().info("Invalid xarm wrist cmd %s" %
    msg.data)
        print("invalid wrist cmd ", msg.data)
        return
    # set limits
    newArmPos = float(min(90.0,newArmPos))
    newArmPos = float(max(-90.0,newArmPos))
    self.arm.setPosition(2,newArmPos, True)
```

This function call is executed when the xarmWrist topic is published. This command just moves the wrist rotation, which we would use to align the fingers of the hand to the object we are picking up. I added some exception handling for invalid values, along with limit checking on the range of the input, which I consider a standard practice for external inputs. We don't want the arm to do something weird with an invalid input, such as if someone was able to send a string on the xarmWrist topic instead of an integer. We also check whether the range of the data in the command is valid, which in this case is from 0 to 1000 servo units. If we get an out-of-bounds error, we clamp the command to the allowed range using the min and max functions.

16. The end effector command and base command (which controls the left-right rotation of the entire arm) work exactly the same way:

```
def effCallback(self, msg):
# set just the end effector position
    try:
        newArmPos = int(msg.data)
    except ValueError:
        self.get_logger().info("Invalid xarm effector cmd
%s" % msg.data)
        return
    # set limits
    newArmPos = min(1000,newArmPos)
    newArmPos = max(0,newArmPos)
    self.arm.setPosition(1,newArmPos)

def baseCallback(self, msg):
# set just the base azimuth position
    try:
        newArmPos = int(msg.data)
    except ValueError:
        self.get_logger().info("Invalid xarm base cmd %s" %
msg.data)
        return
    # set limits
    newArmPos = min(1000,newArmPos)
    newArmPos = max(0,newArmPos)
    self.arm.setPosition(6,newArmPos)
```

The setArm command lets us send one comment to set the position of every servo motor at once, with one command. We send an array of six integers, and this program relays that to the servo motor controller.

As mentioned before, I put in a special value, 9999, that tells this bit of code to not move that motor. This lets us send commands to the arm that move some of the servos, or just one of them. This lets us move the up/down axis and left/right axis of the end of the arm independently, which is important.

Another thing that is important is that while this bit of Python code executes almost instantly, the servo motors take a finite amount of time to move. We have to throw in some delays between the servo commands so that the servo controller can process them and send them to the right motor. I've discovered that the value 0.1 (1/10 of a second) between commands works. If you leave this value out, only one servo will move, and the arm will not process the rest of the commands. The servos use a serial interface in a daisy chain fashion, which means they relay messages to each other. Each servo is plugged into one other servo, which is a big improvement over all the servos being plugged in individually.

17. We can finish up our arm control code with MAIN – the executable part of the program:

```
######################MAIN######################################
rclpy.init()
print("Arm Control Active")
xarmCtr = xarmControl()
# spin ROS 2
rclpy.spin(xarmCtr)
# destroy node explicitly
xarmCtr.destroy_node()
rclpy.shutdown()
```

Here, we initialize rclpy (ROS 2 Python interface) to connect our program to the ROS infrastructure. Then we create an instance of our xarm control class we created. We'll call it xarmCtr. Then we just have to tell ROS 2 to execute. We don't even need a loop. The program will perform publish and subscribe calls, and our timer sends out telemetry, which is all included in our xarmControl object. When we fall out of spin, we are done with the program, so we shut down the ROS node, and then the program.

Now we are ready to start training our robot arm! To do this, we are going to use three different methods to train our arm to pick up objects. In the first stage, we will just train the robot hand – the end effector – to grasp objects. We will use Q-learning, a type of RL, to accomplish this. We will have the robot try to pick up items, and we will reward, or give points, if the robot is successful and subtract points if it fails. The software will try to maximize the reward to get the most points, just like playing a game. We will generate different policies, or action plans, to make this happen.

Introducing Q-learning for grasping objects

Training a robot arm end effector to pick up an oddly shaped object using the **Q-learning** RL technique involves several steps. Here's a step-by-step explanation of the process:

1. Define the state space and action space:

   - **Define the state space**: This includes all the relevant information about the environment and the robot arm, such as the position and orientation of the object, the position and orientation of the end effector, and any other relevant sensor data

   - **Define the action space**: These are the possible actions the robot arm can take, such as rotating the end effector, moving it in different directions, or adjusting its gripper

2. **Set up the Q-table**: Create a Q-table that represents the state-action pairs and initialize it with random values. The Q-table will have a row for each state and a column for each action. As we test each position that the arm moves to, we will store the reward that was computed by the Q-learning equation (introduced in the *Machine learning for robot arms* section) in this table so that we can refer to it later. We will search the Q-table by state and action to see which state-action pair results in the largest reward.

3. **Define the reward function**: Define a reward function that provides feedback to the robot arm based on its actions. The reward function should encourage the arm to pick up the object successfully and discourage undesirable behavior.

4. **Start the training loop**: Start the training loop, which consists of multiple episodes. Each episode represents one iteration of the training process:

   - Reset the environment and set the initial state

   - Select an action based on the current state using an exploration-exploitation strategy such as epsilon-greedy, where you explore random actions with a certain probability (epsilon) or choose the action with the highest Q-value

   - Execute the selected action and observe the new state and the reward

   - Update the Q-value in the Q-table using the Q-learning update equation, which incorporates the reward, the maximum Q-value for the next state, and the learning rate (alpha) and discount factor (gamma) parameters

   - Update the current state to the new state

   - Repeat the previous steps until the episode terminates, either by successfully picking up the object or reaching a maximum number of steps

5. **Exploration and exploitation**: Adjust the exploration rate (represented by epsilon) over time to gradually reduce exploration and favor exploitation of the learned knowledge. This allows the robot arm to initially explore different actions and gradually focus on exploiting the learned information to improve performance.

6. **Repeat training**: Continue the training loop for multiple episodes until the Q-values converge or the performance reaches a satisfactory level.

7. **Perform testing**: After training, use the learned Q-values to make decisions on the actions to take in a testing environment. Apply the trained policy to the robot arm end effector, allowing it to pick up the oddly shaped object based on the learned knowledge.

Note

Implementing Q-learning for training a robot arm end effector requires a combination of software and hardware components, such as simulation environments, robotic arm controllers, and sensory input interfaces. The specifics of the implementation can vary depending on the robot arm platform and the tools and libraries being used.

Writing the code

Now we'll implement the seven-step process we just described by building the code that will train the arm, using the robot arm interface we made in the previous section:

1. First, we include our imports – the functions we'll need to implement our training code:

```
import rclpy
import time
import random
from rclpy.node import Node
from std_msgs.msg import String, Int32MultiArray, Int32
from sensor_msgs.msg import Image
from vision_msgs.msg import Detection2D
from vision_msgs.msg import ObjectHypothesisWithPose
from vision_msgs.msg import Detection2DArray
import math
import pickle
```

   rclpy is the ROS 2 Python interface. We use Detection2D to talk to the vision system from the previous chapter (YOLOV8). I'll explain the pickle reference when we get to it.

2. Next, let's define some functions we'll be using later:

```
global learningRate = 0.1 # learning rate
def round4(x):
 return (math.round(x*4)/4)
# function to restrict a variable to a range. if x < minx, x=min
x,etc.
def rangeMinMax(x,minx,maxx):
 xx = max(minx,x)
 xx = min(maxx,xx)
 return xx
def sortByQ(listByAspect):
 return(listByAspect[2])
```

The learning rate is used in reinforcement learning just like in other machine learning algorithms to adjust how fast the system makes changes as a result of inputs. We'll start with 0.1. If this value is too big, we will have big jumps in our training that can cause erratic outputs. If it's too small, we'll have to do a lot of repetitions. actionSpace is the list of possible hand actions that we are teaching. These values are the angle of the wrist in degrees. Note that -90 and +90 are the same as far as grasping is concerned.

The round4 function is used to round off the aspect ratio of the bounding box. When we detect a toy, as you may remember, the object recognition system draws a box around it. We use that bounding box as a clue to how the toy is oriented relative to the robot. We want a limited number of aspect angles to train for, so we'll round this off to the nearest 0.25.

The SortbyQ function is a custom sort key that we'll use to sort our training to put the highest reward – represented by the letter Q – first.

3. In this step, we'll declare the class that will teach the robot to grasp objects. We'll call the class LearningHand, and we'll make it a node in ROS 2:

```
class LearningHand(Node):
    def __init__(self):
        super().__init__('armQLearn') # node name
        # we need to both publish and subscribe to the RobotCmd
topic
        self.armPosSub = self.create_subscription(Int32MultiArray,
"xarm_pos", self.armPosCallback, 10)
        self.cmdSubscribe = self.create_subscription(String,
'RobotCmd', self.cmdCallback,10)
        self.cmdPub = self.create_publisher(String, 'RobotCmd',
10)
        self.wristPub = self.create_publisher(Int32,'xarmWrist', 10)
        # declare parameter for number of repetitions
        self.declare_parameter('ArmLearningRepeats', rclpy.Parameter.
Type.INTEGER)
        # get the current value from configuration
        self.repeats = self.get_parameter('ArmLearningRepeats').get_
parameter_value().int_value
```

Here we initialize the object by passing up the init function to the parent class (with super). We give the node the name armQLearn, which is how the rest of the robot will find it.

Our ROS interface subscribes to several topics. We need to talk to the robot arm, so we subscribe to xarm_pos (arm position). We need to subscribe (like every program that talks to the robot) to RobotCmd, which is our master mode command channel. We also need to be able to send commands on RobotCmd, so we create a publisher on that topic. Finally, we use a ROS parameter to set the value of how many repetitions we want for each learning task.

4. This next block of code completes the setup for the learning function:

```
self.mode = "idle"
self.armInterface = ArmInterface()
# define the state space
self.stateActionPairs = []
# state space is the target aspect and the hand angle
# aspect is length / width length along x axis(front back)
width on y axis)
aspects = [0.25, 0.5, 0.75, 1, 1.25, 1.5, 1.75]
handAngles = [90, -45, 0, 45] # note +90 and -90 are the same
angle
for jj in range(0,len(aspects)):
```

```
    for ii in range(0,4):
        self.stateActionPairs.append([aspects[jj],
handAngles[ii],0.0])
```

We set the learning system mode to `idle`, which just means "wait for the user to start learning." We create the arm interface by instantiating the `ArmInterface` class object we imported. Next, we need to set up our learning matrix, which stores the possible aspects (things we can see) and the possible actions (things we can do). The last element, which we set to 0 here, is the Q value, which is where we store our training results.

5. The following set of functions helps us to command the arm:

```
def sndCmd(self,msgStr):
    msg = String()
    msg.data = msgStr
    self.cmpPub.publish(msg)

def setHandAngle(self,ang):
    msg = Int32()
    msg.data = ang
    self.wristPub.publish(msg)

def armPosCallback(self,msg):
    self.currentArmPos = msg.data

def setActionPairs(self,pairs):
    self.stateActionPairs = pairs
```

`sndCmd` (send command) publishes on the `RobotCmd` topic and sets arm modes. `SetHandAngle`, as you expect, sets the angle of the wrist servo. `armPosCallback` receives the arm's current position, which is published by the arm control program. `setActionPairs` allows us to create new action pairs to learn.

6. Now we are ready to do the arm training. This is a combined human and robot activity, and is really a lot of fun to do. We'll try the same aspect 20 times:

```
def training(self, aspect):
  # get the aspect from the vision system
  #aspect = 1.0 # start here

  stateActionPairs.sort(key=sortByQ) # sort by Q value
  if len(stateActionPairs)<1:
    #error - no aspects found!
    #
    self.get_logger().error("qLearningHand No Aspect for
    Training")
```

```
        return
    else:
        mySetup = stateActionPairs[0] # using the highest q value
        handAngle = mySetup[1]
        myOldQ = mySetup[2]
```

This initiates the training program on the robot arm. We start by training based on aspect. We first look at our `stateActionPairs` to sort on the highest Q value for this aspect. We use our custom `SortbyQ` function to sort the list of `stateActionPairs`. We set the hand angle to the angle with the highest Q, or expected reward.

7. This part of the program is the physical motion the robot arm will go through:

```
        sndCmd("ARM MID_CARRY")
        timer.pause(1.0)
        sndCmd("ARM GRASP")
        time.sleep(1.0)
        setHandAngle(handAngle)
        time.sleep(0.3)
        # close the gripper
        sndCmd("ARM GRASP_CLOSE")
        time.sleep(0.5)
        # now raise the arm
        sndCmd("ARM MID_CARRY")
        time.sleep(1.0)
```

We start by telling the arm to move to the *Mid Carry* position – halfway up. Then we wait 1 second for the arm to complete its motion, and then we move the arm to the grasp position. The next step moves the wrist to the angle that we got from the Q function. Then we close the gripper with the `ARM GRASP_CLOSE` command. Now we raise the arm to see whether the gripper can lift the toy, using the `ARM MID_CARRY` instruction. If we are successful, the robot arm will now be holding toy. If not, the gripper will be empty.

8. Now we can check to see whether the gripper has an object in it:

```
        #check to see if grip is OK
        handPos = self.currentArmPos[0]
        gripSuccess = False
        if handPos > 650: ## fail
                gripSuccess = -1 # reward value of not gripping
        else: # success!
                gripSuccess = +1 # reward value of gripping
```

If the grip of the robot hand is correct, the toy will prevent the gripper from closing. We check the hand position (which the arm sends twice a second) to see the position. For my particular arm, the position that corresponds to 650 servo units or greater is completely closed. Your arm may vary, so check to see what the arm reports for a fully closed and empty gripper. We set the `gripSuccess` variable as appropriate.

9. Now we do the machine learning part. We use my special modified Bellman equation introduced in the *Machine learning for robot arms* section to adjust the Q value for this state-action pair:

```
# the Bellman Equation
### Q(s, a) = Q(s, a) + α * [R + γ * max(Q(s', a')) - Q(s, a)]
newQ = myOldQ + (learningRate*(gripSuccess))
mySetup[2]=newQ
```

Since we are not using a future reward value (we get the complete reward from this one action of closing the gripper and raising the arm), we don't need the expected future reward, only a present reward. We multiply the `gripSuccess` value (+1 or -1) by the learning rate and add this to the old Q score to get a new Q score. Each success increments the reward while any failure leads to a decrement.

10. To finish our learning function, we insert the updated Q value back into the learning table that matches the aspect angle and the wrist angle we tested:

```
foundStateActionPair = False
# re insert back into q learning array
for i in range (0,len(stateActionPairs):
    thisStateAction = stateActionPairs[i]
    if thisStateAction[0] == mySetup[0] and
thisStateAction[1] == mySetup[0]:
        foundStateActionPair=True
        stateActionPairs[2]=mySetup[2] # store the new q value
in the table
    if not foundStateActionPair:
        # we don't have this in the table - let's add it
        stateActionPairs.append(mySetup)
input("Reset and Press Enter") # wait for enter key to continue
```

If this state-action pair is not in the table (which it should be), then we add it. I put this in just to keep the program from erroring out if we give a strange arm angle. Finally, we pause the program and wait for the user to hit the *Enter* key in order to continue.

11. Let's now look at the rest of the program, which is pretty straightforward. We have to do some housekeeping, service some calls, and make our main training loop:

```
def cmdCallBack(self,msg):
  robotCmd = msg.data
  if robotCmd == "GoLearnHand":
```

```
      self.mode = "start"
  if robotCmd == "StopLearnHand":
      self.mode = "idle"
```

This cmdCallBack receives commands from the RobotCmd topic. The only two commands we service in this program are GoLearnHand, which starts the learning process, and StopLearnHand, which lets you stop training.

12. This section is our arm interface to the robot arm and sets up the publish/subscribe interface we need to command the arm:

```
class ArmInterface():
 init(self):
    self.armPosSub = self.create_subscription(Int32MultiArray,
'xarm_pos',self.armPosCallback, 10)
    self.armAngSub = self.create_subscription(Int32MultiArray,
'xarm_angle',self.armAngCallback, 10)
    self.armPosPub = self.create_publisher(Int32MultiArray,
'xarm')

 def armPosCallback(self,msg):
    self.armPos = msg.data

 def armAngCallback(self, msg):
    self.armAngle = msg.data
    # decoder ring: [grip, wrist angle, wrist pitch, elbow pitch,
  sholder pitch, sholder yaw]

 def setArmPos(self,armPosArray):
    msg = Int32MultiArray
    msg.data = armPosArray
    self.armPosPub.publish(msg)
```

We subscribe to xarm_pos (arm position in servo units) and xarm_angle (arm position in degrees). I added the ability to set the robot arm position on the xarm topic, but you may not need that.

For each subscription we need a callback function. We have armPosCallback and armAngleCallback, which will be called when the arm publishes its position, which I set to happen at 2 Hertz, or twice a second. You can increase this rate in the xarm_mgr program if you feel it necessary.

13. Now we get to the main program. For a lot of ROS programs, this main section is pretty brief. We have an extra routine we need to put here. To save the training function after we do our training, I came up with this solution – to *pickle* the state-action pairs and put them into a file:

```
### MAIN ####
# persistent training file to opeate the arm
ArmTrainingFileName = "armTrainingFile.txt"
armIf = ArmInterface()
armTrainer = LearningHand()
#open and read the file after the appending:
try:
 f = open(ArmTrainingFileName, "r")
 savedActionPairs = pickle.load(f)
 armTrainer.setActionPairs(savedActionPairs)
 f.close()
except:
 print("No Training file found")
 self.get_logger().error("qLearningHand No Training File Found
armTrainingFile.txt")
```

When we run this program, we need to load this file and set our action-pairs table to these saved values. I set up a try/except block to send an error message when this training file is not found. This will happen the first time you run the program, but we'll create a new file in just a moment for next time.

We also instantiate our class variables for the arm trainer and the arm interface, which creates the main part of our training program.

14. This is the meat of our training loop. We set the aspect and number of trial repetitions that we train on:

```
aspectTest = [1.0, 0.5, 1.5,2]
trainingKnt = 20
for jj in aspectTest:
 for ii in range(0,trainingKnt):
   print("Starting Training on Aspect ", jj)
   armTrainer.training(jj)
```

Start with the toy parallel to the front of the robot. Do 20 trials of picking it up, and then move the toy 45 degrees to the right for the next part. Then perform 20 more trials. Then the toy is moved to be 90 degrees to the robot. Run 20 trials. Finally set the toy at -45 degrees (to the left) for the final set and run 20 times. Welcome to machine learning!

15. You'll probably guess that the last thing we do is save our training data, like this:

```
f = open("ArmTrainingFileName", "w")
# open file in write mode
pickle.dump(armTrainer.stateActionPairs,f)
print("Arm Training File Written")
f.close()
```

This completes our training program. Repeat this training for as many types of toys as you have, and you should have a trained arm that consistently picks up toys at a variety of angles to the robot. Start with a selection of toys you want the robot to pick up. Set the angle of the toy to the robot at 0 – let's say this is with the longest part of the toy parallel to the front of the robot. Then we send GoLearnHand on RobotCmd to put the robot arm in learning mode.

We have tried out Q-learning in a couple of different configurations, with a limited amount of success in training our robot. The main problem with Q-learning is that we have a very large number of possible states, or positions, that the robot arm can be in. This means that gaining a lot of knowledge about any one position by repeated trials is very difficult. Next, we are going to introduce a different approach using GAs to generate our movement actions.

Introducing GAs

Moving the robot arm requires the coordination of three motors simultaneously to create a smooth movement. We need a mechanism to create different combinations of motor movement for the robot to test. We could just use random numbers, but that would be inefficient and could take thousands of trials to get to the level of training we want.

What if we had a way of trying different combinations of motor movement, and then pitting them against one another to pick the best one? It would be a sort of Darwinian *survival of the fittest* for arm movement scripts – such as a GA process. Let's explore how we can apply this concept to our use case.

Understanding how the GA process works

Here are the steps involved in our GA process:

1. We do a trial run to go from position 1 (neutral carry) to position 2 (pickup). The robot moves the arm 100 times before getting the hand into the right position. Why 100? We need a large enough sample space to allow the algorithm to explore different solutions. With a value of 50, the solution did not converge satisfactorily, while a value of 200 yielded the same result as 100.

2. We score each movement based on the percentage of goal accomplishment, indicating how much this movement contributed to the goal.

3. We take the 10 best moves and put them in a database.

4. We run the test again and do the same thing – now we have 10 more *best moves* and 20 moves in the database.

5. We take the five best from the first set and cross them with the five best from the second set – plus five moves chosen at random and five more made up of totally random moves. Crossing two solutions refers to the process of taking a segment from the first set and a segment from the second set. In genetic terms, this is like taking half the *DNA* from each of two *parents* to make a new *child*.

6. We run that sequence of moves, and then take the 10 best individual moves and continue on.

Through the process of selection, we should quickly get down to a sequence that performs the task. It may not be optimal, but it will work. We are managing our *gene pool* (a list of trial solutions to our problem) to create a solution to a problem by successive approximation. We want to keep a good mix of possibilities that can be combined in different ways to solve the problem of moving our arm to its goal.

We can actually use several methods of **cross-breeding** our movement sequences. What I described is a simple cross – half the first parent's genetic material and half the second parent's material (if you will pardon the biological metaphor). We could instead use quarters – ¼ first, ¼ second, ¼ first, ¼ second – to have two crosses. We could also randomly grab bits from one or the other. We will stick with the half/half strategy for now, but you are free to experiment to your heart's content. In essence, in all of these options, we are taking a solution, breaking it in half, and randomly combining it with half of a solution from another trial.

You are about to issue an objection: what if the movement takes less than 10 steps? Easy – when we get to the goal, we stop, and discard the remaining steps.

> **Note**
> We are not looking for a perfect or optimum task execution, but just something good enough to get the job done. For a lot of real-time robotics, we don't have the luxury of time to create a perfect solution, so any solution that gets the job done is adequate.

Why did we add the five additional random sample moves, and five totally random moves? This also mimics natural selection – the power of mutation. Our genetic code (the DNA in our bodies) is not perfect, and sometimes inferior material gets passed along. We also experience random mutations from bad copies of genes, cosmic rays, and viruses. We are introducing some random factors to *bump* the tuning of our algorithm – the element of natural selection – in case we converge on a local minimum or miss some simple path because it has not occurred yet to our previous movements.

But why on Earth are we going to all this trouble? The GA process can do something very difficult for a piece of software – it can innovate or evolve new solutions out of primitive actions by basically trying stuff until it finds out what works and what does not. We have provided another machine learning process to add to our toolbox, but one that can create solutions we, the programmers, had not preconceived.

Now, let's dive into the GA process. In the interest of transparency, we are going to build our own GA process from scratch.

> **Note**
>
> We'll be building our own tools in this version, but there are some prebuilt toolsets that can help you to create GAs, such as **Distributed Evolutionary Algorithms in Python** (DEAP), which can be found at `https://github.com/DEAP` and installed by typing in `pip install deap`.

Building a GA process

We loosely adopt the concept of the *survival of the fittest* to decide which plans are the fittest and get to survive and propagate. I'm giving you a sandbox in which to play genetic engineer, where you have access to all of the parts and nothing is hidden behind the curtain. You will find that for our problem, the code is not all that complex:

1. We'll start by creating the `computefitness` function, the one that scores our genetic material. **Fitness** is our criteria for grading our algorithm. We can change the fitness to our heart's content to tailor our output to our needs. In this case, we are making a path in space for the robot arm from the starting location to the ending goal location. We evaluate our path in terms of how close any point of the path comes to our goal. Just as in our previous programs, the movement of the robot is constituted as 27 combinations of the three motors going clockwise, counterclockwise, or not moving. We divide the movement into small steps, each three motor units (1.8 degrees) of so of motion. We string together a whole group of these steps to make a path. The fitness function steps along the path and computes the hand position at each step.

2. The `predictReward` function makes a trial computation of where the robot hand has moved as a result of that step. Let's say we move *Motor 1* clockwise three steps, leave *Motor 2* alone, and move *Motor 3* counterclockwise three steps. This causes the hand to move slightly up and out. We score each step individually by how close it comes to the goal. Our score is computed out of 100; 100 is exactly at the goal, and we take away one point for each 1/100th of the distance the arm is away from the goal, up to a maximum of 340 mm. Why 340? That is the total length of the arm. We score the total movement a bit differently than you might think. Totaling up the rewards make no difference, as we want the point of closest approach to the goal. So we pick the single step with the highest reward and save that value. We throw away any steps after that, since they will only take us further away. Thus we automatically prune our paths to end at the goal.

3. I used the term `allele` to indicate a single step out of the total path, which I called `chrom`, short for chromosome:

```
def computeFitness(population, goal, learningRate, initialPos):
  fitness = []
  gamma = 0.6
  state=initialPos
  index = 0
  for chrom in population:
    value=0
    for allele in chrom:
      action = ACTIONMAT[allele]
      indivFit, state =
      predictReward(state,goal,action,learningRate) value +=
      indivFit
      if indivFit > 95:
        # we are at the goal - snip the DNA here
        break
    fitness.append([value,index])
    index += 1
  return fitness
```

4. How do we create our paths to start with? The `make_new_individual` function builds our initial population of chromosomes, or paths, out of random numbers. Each chromosome contains a path made up of a number from 0 to 26 that represents all the valid combinations of motor commands. We set the path length to be a random number from 10 to 60:

```
def make_new_individual():
  # individual length of steps
  lenInd = random.randint(10,60)
  chrom = [] # chromosome description
  for ii in range(lenInd):
    chrom.append(randint(26))
  return chrom
```

5. We use the `roulette` function to pick a portion of our population to continue. Each generation, we select from the top 50% of scoring individuals to donate their DNA to create the next generation. We want the reward value of the path, or chromosome, to weigh the selection process; the higher the reward score, the better chance of having children. This is part of our selection process:

```
# select an individual in proportion to its value
def roulette(items):
  total_weight = sum(item[0]
  for item in items)
```

```
    weight_to_target = random.uniform(0, total_weight)
    for item in items:
     weight_to_target -= item[0]
     if weight_to_target <= 0:
      return item
# main Program
INITIAL_POS = [127,127,127]
GOAL=[-107.39209423, -35.18324771]
robotArm=RobotArm()
robotArm.setGoal(GOAL)
population = 300
learningRate = 3
crossover_chance = .50
mutate_chance = .001
pop = []
```

6. We start by building our initial population out of random parts. Their original fitness will be very low: about 13% or less. We maintain a pool of 300 individual paths, which we call chromosomes:

```
for i in range(population): pop.append(make_new_individual())
    trainingData=[] epochs = 100
```

7. Here we set up the loop to go through 100 generations of our natural selection process. We begin by computing the fitness of each individual and adding that score to a fitness list with an index pointing back to the chromosome:

```
for jj in range(epochs):
    # evaluate the population
    fitnessList = computeFitness(pop,GOAL,learningRate, INITIAL_
POS)
```

8. We sort the fitness in inverse order to get the best individuals. The largest number should be first:

```
fitnessList.sort(reverse=True)
```

9. We keep the top 50% of the population and discard the bottom 50%. The bottom half is out of the gene pool as being unfit:

```
fitLen = 150
fitnessList = fitnessList[0:fitLen] # survival of the fittest...
```

10. We pull out the top performer from the whole list and put it into the **hall of fame** (**HOF**). This will eventually be the output of our process. In the meantime, we use the HOF or **HOF fitness** (**HOFF**) value as a measure of the fitness of this generation:

```
hoff = pop[fitnessList[0][1]]
print("HOF = ",fitnessList[0])
```

11. We store the HOFF value in a `trainingData` list so we can graph the results at the end of the program:

```
trainingData.append(fitnessList[0][0])
newPop = []
for ddex in fitnessList: newPop.append(pop[ddex[1]])
  print ("Survivors: ",len(newPop))
```

12. At this phase, we have deleted the bottom 50% of our population, removing the worst performers. Now we need to replace them with the children of the best performers of this generation. We are going to use crossover as our mating technique. There are several types of genetic mating that can produce successful offspring. Crossover is popular and a good place to start, as well as being easy to code. All we are doing is picking a spot in the genome and taking the first half from one parent, and the second half from the other. We pick our parents to *mate* randomly from the remaining population, weighted proportionally to their fitness. This is referred to as **roulette wheel selection**. The better individuals are weighted more heavily and are more likely to be selected for breeding. We create 140 new individuals as children of this generation:

```
# crossover
# pick to individuals at random # on the basis of fitness
numCross = population-len(newPop)-10 print ("New Pop
Crossovers",numCross) # #
# add 5 new random individuals for kk in range(10):
newPop.append(make_new_individual())
for kk in range(int(numCross)):
 p1 = roulette(fitnessList)[1]
 p2 = roulette(fitnessList)[1]
 chrom1 = pop[p1]
 chrom2 = pop[p2]
 lenChrom = min(len(chrom1),len(chrom2)) xover =
 randint(lenChrom)
 # xover is the point where the chromosomes cross over newChrom
 = chrom1[0:xover]+chrom2[xover:]
```

13. Our next step is **mutation**. In real natural selection, there is a small chance that DNA will get corrupted or changed by cosmic rays, miscopying of the sequence, or other factors. Some mutations are beneficial, and some are not. We create our version of this process by having a small chance (1/100 or so) that one gene in our new child path is randomly changed into some other value:

```
# now we do mutation bitDex = 0
for kk in range(len(newChrom)-1):
  mutDraw = random.random()
  if mutDraw < mutate_chance: # a mutation has occured!
  bit = randint(26)
```

```
        newChrom[kk]=bit
        print ("mutation")
   newPop.append(newChrom)
```

14. Now that we have done all our processing, we add this new child path to our population, and get ready for the next generation to be evaluated. We record some data and loop back to the start:

```
# welcome the new baby from parent 1 (p1) and parent 2 (p2)
print("Generation: ",jj,"New population = ",len(newPop))
pop=newPop
mp.plot(trainingData) mp.show()
```

So, how did we do with our mad genetic experiment? The following output chart speaks for itself:

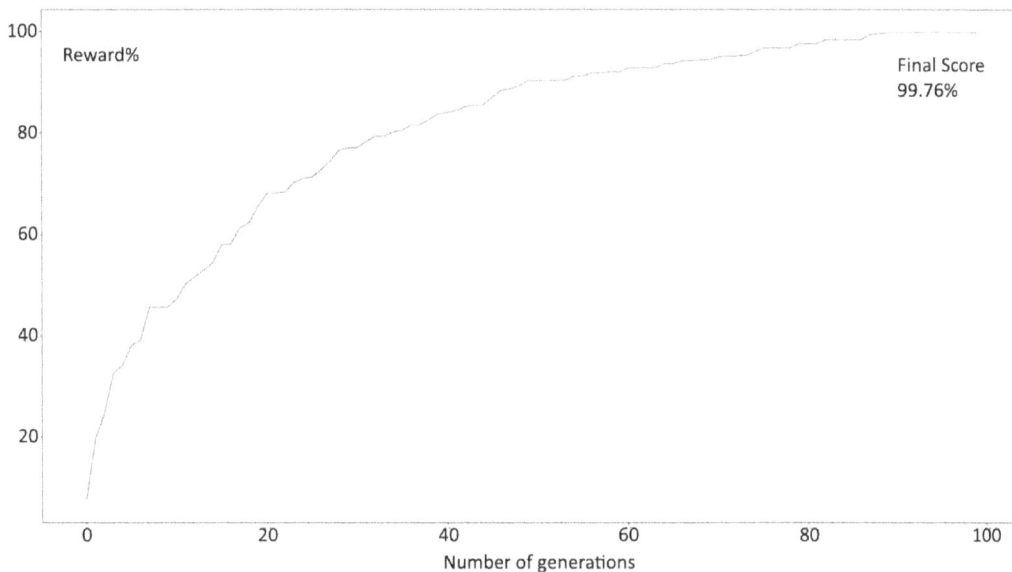

Figure 5.5 – Learning curve for the GA solution

The GA, for all it seems like a bit of voodoo programming, works quite well as a machine learning tool for this specific case of training our robot arm. Our solution peaked at 99.76% of the goal (about 2 mm) after just 90 generations or so, which is quite fast for an AI learning process. You can see the smooth nature of the learning that shows that this approach can be used to solve path-planning problems for our robot arm. I have to admit that I was quite skeptical about this process, but it seems to work quite well for this particular problem domain.

The programming really was not too hard, and you can spend some time improving the process by tweaking the parameters of the GA. What if we had a smaller population? What if we changed the fitness criteria? Get in there, muck about, and see what you can learn.

Alternative robot arm ML approaches

The realm of robot arm control via machine learning is really just getting started. There are a couple of research avenues I wanted to bring to your attention as you look for further study. One way to approach our understanding of robot movement is to consider the balance between *exploitation* and *exploration*. Exploitation is getting the robot to its goal as quickly as possible. Exploration is using the space around the robot to try new things. The path-planning program may have been stuck on a local minimum (think of this as a blind alley), and there could be better, more optimal solutions available that had not been considered.

There is also more than one way to teach a robot. We have been using a form of self-exploration in our training. What if we could show the robot what to do and have it learn by example? We could let the robot observe a human doing the same task, and have it try to emulate the results. Let's discuss some alternative methods in the following sections.

Google's SAC-X

Google is trying a slightly different approach to the robot arm problem. In their **Scheduled Auxiliary Control** (**SAC- X**) program, they surmise that it can be quite difficult to assign reward points to individual movements of the robot arm. They break down a complex task into smaller auxiliary tasks, and give reward points for those supporting tasks to let the robot build up to a complicated challenge. If we were stacking blocks with a robot arm, we might separate picking up the block as one task, moving with the block in hand as another, and so on. Google referred to this as a *sparse reward* problem if reinforcement was only used on the main task, stacking a block on top of another. You can imagine how, in the process of teaching a robot to stack blocks, there would be thousands of failed attempts before a successful move resulted in a reward.

Amazon Robotics Challenge

Amazon has millions and millions of boxes, parts, bits, and other things on its shelves. The company needs to get the stuff from the shelves into small boxes so they can ship it to you as fast as possible when you order it. For the last few years, Amazon has sponsored the *Amazon Robotics Challenge*, where teams from universities were invited to use robot arms to pick up items off a shelf and, you guessed it, put them into a box.

When you consider that Amazon sells almost everything imaginable, this is a real challenge. In 2017, a team from Queensland, Australia, won the challenge with a low-cost arm and a really good hand-tracking system.

Summary

Our task for this chapter was to use machine learning to teach the robot how to use its robot arm. We used two techniques with some variations. We used a variety of reinforcement learning techniques, or Q-learning, to develop a movement path by selecting individual actions based on the robot's arm state. Each motion was scored individually as a reward, and as part of the overall path as a value. The process stored the results of the learning in a Q-matrix that could be used to generate a path. We improved our first cut of the reinforcement learning program by indexing, or encoding, the motions from a 27-element array of possible combinations of motors as numbers from 0 to 26, and likewise indexing the robot state to a state lookup table. This resulted in a 40x speedup of the learning process. Our Q-learning approach struggled with the large number of states that the robot arm could be in.

Our second technique was a GA. We created individual random paths to make a population. We created a fitness function to score each path against our goal and kept the top performers from each generation. We then crossed genetic material from two somewhat randomly selected individuals to create a new child path. The GA also simulated mutation by having a slight chance of random changes in the steps of a path. The results for the GA showed no problem with the state space complexity of our robot arm and generated a valid path after just a few generations.

Why do we go to all of this trouble? We use machine learning techniques when other empirical methods are either difficult, not reliable, or don't produce solutions in a reasonable amount of time. We can also tackle much more complex tasks with these techniques that might be intractable to a brute-force or math-only solution.

In the next chapter, we'll be adding a voice interface to the robot with natural language processing, so you can talk to the robot and it will listen – and talk back.

Questions

1. In Q-learning, what does the Q stand for?

    Hint: You will have to research this yourself.

2. What could we do to limit the number of states that the Q-learning algorithm has to search through?

3. What effect does changing the learning rate have on the learning process?

4. What function or parameter serves to penalize longer paths in the Q-learning equation? What effect does increasing or decreasing this function have?

5. In the genetic algorithm, how would you go about penalizing longer paths so that shorter paths (fewer number of steps) would be preferred?

6. Look up the SARSA variation of Q-learning. How would you implement the SARSA technique into program 2.

7. What effect does changing the learning rate in the genetic algorithm have? What are the upper and lower bounds of the learning rate?

8. In a genetic algorithm, what effect does reducing the population have?

Further reading

- *Python Deep Learning* by Zocca, Spacagna, Slater, and Roelants, Packt Publishing

- *Artificial Intelligence with Python* by Prateek Joshi, Packt Publishing

- *AI Junkie: Genetic Algorithm – A Brief Overview*, retrieved from `http://www.ai-junkie.com/ga/intro/gat2.html`

- *Basic Reinforcement Learning Tutorial 2: SARSA*: `https://github.com/vmayoral/basic_reinforcement_learning/tree/master/tutorial2`

- *Google DeepMind Blog: Learning by Playing (Robot Arm (SAC-X))*: `https://deepmind.com/blog/learning-playing/`

6

Teaching a Robot to Listen

Teaching a robot to listen to spoken instructions is a whole discipline in itself. It is not sufficient for the robot to just recognize individual words or some canned phrase. We want the robot to respond to normal spoken commands with a normal variety of phrasing. We might say, "Pick up the toys," or "Please pick up all the toys," or "Clean this mess up," any of which would be a valid command to instruct the robot to begin searching the room for toys to pick up and put away. We will be using a variety of techniques and processes for this chapter. We are going to be building on an open source verbal assistant called **Mycroft**, an AI-based speech recognition and **natural language processing** (**NLP**) engine that can be programmed and extended by us. We will be adding some additional capability to Mycroft – we will use a technique I call the "fill in the blank" method of command processing to extract the intent of the user's voice instructions, so that the robot does what you want it to do, even if that is not exactly what you said. We will complete this chapter by teaching the robot to both tell and respond to a specific form of human communication – knock-knock jokes.

The following topics will be covered in the chapter:

- Exploring robot speech recognition with NLP – both **speech to text** (**STT**) and **text to speech** (**TTS**)

- Programming our robot

Technical requirements

This chapter uses the following tools:

- Mycroft Open Source Voice Assistant (`http://mycroft.ai`) – I had to build it from source from the GitHub repository (`https://github.com/MycroftAI`), so expect to do the same to keep it compatible with the **Robot Operating System** (**ROS**) we run the robot with.

- Python 3.2.

- You will need a GitHub account at `https://github.com/`.

- I used a miniature USB speaker and microphone for this project, which worked very well with the Jetson. They can be found at `https://www.amazon.com/gp/product/B08R95XJW8`

The code used in this chapter can be found in the GitHub repository for this book at `https://github.com/PacktPublishing/Artificial-Intelligence-for-Robotics-2e`.

Exploring robot speech recognition with NLP

This is going to be a rather involved chapter, but all of the concepts are fairly easy to understand. We will end up with a very strong framework to build voice recognition and commands upon. Not only will you get a voice-based command system for a robot, but also a full-featured digital assistant that tells jokes. Let's first quickly introduce NLP.

Briefly introducing the NLP concept

NLP is not just converting sound waves to written words (speech to text, or STT), but also understanding what those words mean. We don't want to just have some rigid, pre-programmed spoken commands, but some ability for the robot to respond to human speech.

We will be using two different forms of STT processing:

- **Spectrum analysis**: This type helps to detect when you say the robot's name. This technique recognizes words or phrases by sampling which frequencies and amplitudes make up the word. This process has the advantage of not taking a lot of computer resources and it is good at recognizing just one word or phrase – our "wake word" that will cause the computer to switch to the second type of voice recognition. This is the reason other voice-operated assistants require you to use a specific word (e.g., Siri, or Alexa) to enable them to start listening.

- **Phoneme recognition**: This technique converts STT by recognizing the parts of sounds – phonemes – that make up words. This technique, which seeks to interpret all sounds as words, is much more difficult, so we use the wake word to trigger the change. We'll cover this in more detail later in this chapter.

Next, let's explore our primary goals for implementing speech recognition.

Setting our goals

We set several goals for our robot in *Chapter 2*, which included being able to give voice commands to the robot since we may be using the robot without a base station. I also wanted the robot to be able to interact with my grandchildren, and specifically to be able to tell and respond to knock-knock jokes, a favorite activity of my grandson, William. For our robot, we do not want to use canned or memorized speech commands but rather have it be able to do some NLP to create a form of robot understanding of the spoken word. For example, if we want to have a command for picking up a toy, we humans could phrase that several ways: *grab a toy*, *grasp a toy*, *pick up that toy car*, or even *get that*. We want the robot to understand or at least respond to all of those utterances with the same action, to drive to the nearest toy and pick it up with the robot arm. STT systems are fairly commonplace today, but we would like to have some natural variations in the robot's speech patterns to help create the illusion that the robot is smarter than it really is. We can break this process down into several steps, which we will be handling independently:

- Receive audio (sound) inputs. We need the robot to be able to hear or have the ability to convert sound into a digital form.

- Those sounds need to be converted into text that the robot can process.

- Use processing on those text words to understand the intent of the speaker. We need to not just recognize individual words but combine those words into sentences and from those sentences, infer the intent of the speaker to understand what the robot is to do.

- Use that intent as a command to perform some task.

- Provide verbal responses in the form of spoken words (text to speech, or TTS) back to the operator to confirm the robot heard and understood the command.

- Create a custom verbal interface that both tells and responds to knock-knock jokes.

We will start in the next section by introducing the process of STT, which is how the robot will receive voice input from you.

Understanding the STT process

In the rest of this chapter, we will be implementing an AI-based voice recognition and response system in the robot and creating our own custom voice interface. We will be using Mycroft, an open source voice-activated digital assistant that is adept at understanding speech and is easily extended for new functions and custom interfaces.

We will discuss each of the steps involved in voice interaction in detail. There are two forms of STT involved in this process that greatly simplify matters for the robot: **wake word recognition** and **STT recognition**. Let's explore wake word recognition first.

Listening for the wake word

In the first approach, the robot is listening continuously for only one sound – the wake word. This is a specific sound that just means one thing – get ready to process the next sound into a command. Why is this necessary? Since the robot has only a very small processor – the Jetson Nano – it really does not have the sort of onboard compute power to run a robust STT engine. But it can run a simple sound recognizer that can listen for the wake word. You are familiar with this from other voice command systems, such as Alexa or Siri, that also either use a special wake word or a button to have the interface pay attention (see https://www.howtogeek.com/427686/how-alexa-listens-for-wake-words/).

Once the wake word is received, the Jetson Nano switches into record mode and records the next thing we say. It then transfers that information to an online system, the Google Cloud Speech to Text system (the same thing that runs the Google Assistant).

How does the robot recognize the wake word? The speech system we will be using, the open source system Mycroft, uses one of two methods:

- The first is a **phoneme recognition system** called Sphynx. What is a phoneme? You can understand that words are made up of individual sounds, which we roughly assign to letters of the alphabet. An example would be the *p* sound in the word *pet* or *pick* – this is an example of a phoneme. The word "Albert" has several phonemes – the *A* sound, (ah), the *L* sound, the *B*, the *ER* together, and finally, the *T*. The letters we associate with the sounds – for example, the *ch* in *cherry*, and the *er* in *Albert*, are called **graphemes**, as they graphically represent these sounds. We could say that the STT problem is one of mapping these phonemes to graphemes, but we know that this is too easy – English has all sorts of borrowed words and phrases where the pronunciation and the spelling are far apart.

  The frontend of the Mycroft speech recognition process uses phonemes to recognize the wake word. You will find that it is quite sensitive. I had no problem getting the speech processor to receive the wake word from eight feet away. When we get to the setup, we will change the default Mycroft wake word from "Hey, Mycroft," to "Hey, Albert."

- Mycroft can also use a trained **neural network** that has been taught to recognize entire words all at once by their **spectral power graph**. What is a spectral graph? The sound of your voice is not one frequency of sound energy – it is a complex congregation of different frequencies produced by our mouths and vocal cords. If we spoke in pure frequencies, we would sound like a flute – pure tones at mostly one frequency. We can use a process called a **fast Fourier transform** to convert a selection of speech into a graph that shows the amount of energy (volume) at each frequency. This is called a spectral plot or spectral graph. The low frequencies are on the left, and higher frequencies are on the right. Most human speech's energy is concentrated in the frequencies between 300 Hz and 4,000 Hz. Each word has a unique distribution of sound energy amounts in these frequencies, and can be recognized by a neural network in this manner:

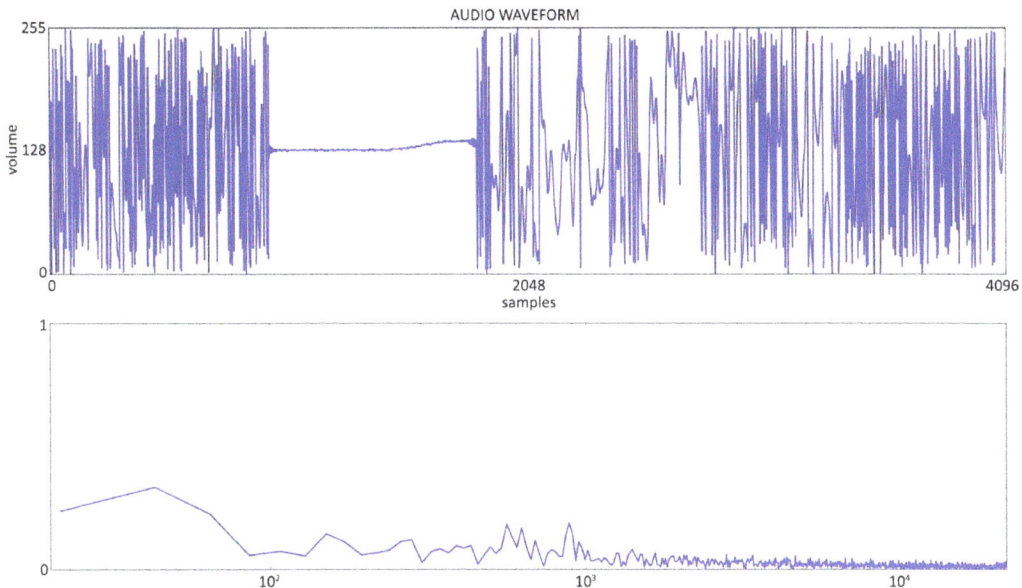

Figure 6.1 – Analog audio waveform (top) and the spectral graph for the phrase "Hey, Albert" (bottom)

This preceding diagram shows both the audio waveform (top graph) in the time domain and the spectral plot in the frequency domain for the phrase "Hey, Albert."

Both the phoneme method and the neural network method use spectral plots to recognize sounds as words, but the phoneme process divides words into individual sounds, and the neural network listens and recognizes the entire word all at once. Why does this make a big difference? The phoneme system can be developed to recognize any word in English without reprogramming or retraining, while the neural network has to be trained on each word individually, and hopefully by a lot of different speakers with a lot of different accents. We'll be using the neural network method for Albert.

Note

You can remember from *Chapter 4* that we needed labeled data to train a neural network. You recall we had pictures in categories and trained on each category. Training **artificial neural networks (ANNs)** for sound is the same: we need sounds and the associated words. Can you think of a place to get samples of lots of different voices where you also have the exact written script to match? Have you ever listened to a book on tape?

Converting STT

Our next step after receiving the wake word is to record the next sounds that the robot hears. The Mycroft system then transfers that audio data over the internet to the Google online STT engine (`https://cloud.google.com/speech-to-text/`). This is a quick way to resolve the problem of our little Jetson Nano not having enough processing power or storage to have a robust speech recognition capability.

What goes on in Google Cloud? The STT engine breaks the speech down into phonemes (sounds) and uses a neural network to assign the most probable graphemes (letters) to those sounds. The output would be spelled out more phonetically than we want to receive. For example, as per the *Carnegie Mellon University Pronouncing Dictionary*, the sentence "Pick up the toys, please?" comes out as `P IH K . AH P . DH AH . T OY Z . P L IY Z`. Why is this the case? What happened? These are the phonemes that make up that sentence. The periods indicate spaces between words. Now the system has to convert this into the words we are expecting. The STT system uses word rules and dictionaries to come up with the most likely conversion into regular words. This includes both expert systems (word rules) and trained neural networks that predict output words based on phonemes.

We can call this step the **language model**. Our STT outputs the sentence "How many ounces in a gallon?" and sends it back to the robot, all in less than two seconds.

Now that we have the command in text, an English sentence, how does the robot recognize what your intent is?

Clarifying the intent

The NLP we are doing has one aim, or goal. We are giving commands to our robot using a voice interface. Commands in English normally follow a sentence pattern, something like "You – do this." Often the "you" subject of the sentence is implied or understood and left out. We are left with statements such as "Clean this room," or "Pick up those toys." The intent of these commands is to have the robot initiate a program that results in the robot picking up toys and putting them away. The robot and its processor have to divine or derive the intent of the user from the words that are spoken. What we want is for any reasonable sentence to have as its meaning: "You, robot, start your pick-up-toys process."

Think of how many ways we can say that command to the robot. Here are some examples:

- Let's clean up this room
- Put away the toys
- Pick up the toys
- Pick up all the toys
- Clean up this room

- Put those away
- Put this away
- Time to clean up

What do these phrases have in common? They all imply the subject who is doing the action is the robot. There are no words such as "You," "robot," or "Albert" to indicate to whom the command is intended for. The word "toys" appears a lot, as does "pick," "clean," and "put away." It is possible that we can just pay attention to those keywords to understand this command. If we get rid of all of the common conjunction and pronoun words, what does the list look like?

- Clean room
- Put toys
- Pick toys
- Pick toys
- Clean room
- Put away
- Put away
- Time clean

An important concept for this chapter is to understand that we are not trying to understand all speech, but only that subset of speech that are commands that the robot can execute. A general solution to this voice recognition problem would be to have some ability to predict from the command given to the robot, the likelihood that the intent of the user points to one command more than any of the others. You can see that in the case of the word "clean," none of our other commands ("drive around," "move arm," or "stop") relate to "clean" at all. Thus, a sentence with "clean" in it is likely associated with the *pick up toys* command. This process of deciding intent will be used later in this chapter to send commands to the robot using Mycroft.

Now we are going to jump right into programming the Albert robot to listen and understand commands using Mycroft.

Programming our robot

As discussed earlier in this chapter, Mycroft is a version of a digital assistant similar to Siri from Apple or Alexa from Amazon in that it can listen to voice commands in a mostly normal fashion and interface those commands to a computer. We are using it because it has an interface that runs on a Jetson Nano 3. In this section, we will be setting up our hardware and our software (i.e., Mycroft).

Setting up the hardware

We will be installing Mycroft on Nvidia Jetson Nano (or whatever microprocessor you're using). One of the few things that the Jetson Nano did not come with is **audio capability**. It has no speakers or microphones. I found that a quick and effective way to add that capability was to use an existing hardware kit that provided both a very high-quality speaker and an excellent set of stereo microphones in a robot-friendly form factor. Note that this works with pretty much any Linux **single-board computer** (SBC).

The kit is a miniature USB audio board that plugs into the Jetson Nano. It has both speakers and a microphone.

> Note
>
> I used a USB audio board (the brand is not important as any of them will do) for the Jetson Nano, which has been working very well for me and fits in the very small space we have on the robot. Installation could not be simpler. Plug in the audio board. You will need to go to **Settings** in the upper-right corner of your screen to select the USB audio version. There will be several other options listed.

Turn on your Jetson Nano 3 with the new speaker and microphone. I ran a quick test with YouTube to make sure the audio worked, and you can test it directly in the **Settings** user interface. Now we can dive into the software.

Setting up the Mycroft software

There are several ways to install Mycroft, as we have to put Mycroft on top of the other software we have already:

1. Since Mycroft must get along with the ROS, and all of the AI libraries we installed, such as TensorFlow, Theano, and Keras, it is best that we use the `git clone` method to download the source code and build Mycroft on the Jetson Nano:

    ```
    git clone https://github.com/MycroftAI/mycroft-core.git cd
    Mycroft-core
    bash dev_setup.sh
    ```

    Mycroft will create a virtual environment it needs to run. It also isolates the Mycroft package from the rest of the packages on the Jetson Nano.

> Note
>
> Please do not install Mycroft as the root user (or superuser). This will cause permissions problems with the configuration files.

2. In order to get the Mycroft system to work in this manner, I also had to do one more step. The Mycroft system kept failing when I first tried to get it to run. It would quit or get stuck when I tried to start the debugger. In order to correct this problem, I had to recompile the entire system using the following steps:

```
sudo rm -R -/.virtualenvs/Mycroft
cd ~/mycroft-core
./dev_setup.sh
```

Once that is done (and it took quite a while – as in several hours), you should be able to run the Mycroft system with the startup commands from the `mycroft-core` directory.

3. You can start in debug mode:

```
./start-mycroft.sh debug
```

Alternatively, you can start in normal mode:

```
./start-mycroft.sh all
```

You will probably be using debug mode quite a bit when you are developing your speech commands. You can't see what's going on without it. Once everything is working properly, you can switch to normal mode, but there will be no user interface except for the robot's voice.

4. Now test that Mycroft is working properly. When you first get Mycroft to run, it will want to be paired with your account on the Mycroft web server. You need to set up a services account on the Mycroft website at `http://home.mycroft.ai`. Then the Jetson Nano will give you a six-letter code to put into the website under **Devices** (on the hamburger menu on the far right-hand side of the website).

    Once the robot is paired with the Mycroft server, it can transfer data back and forth. The wake word will start out being the default, "Hey, Mycroft." You can test that everything is working by first asking, "Hey, Mycroft, what time is it?" Mycroft divides its capabilities into skills that are each controlled by a separate script. The `Time` skill is totally self-contained inside the Jetson Nano. The robot should give you a voice response that is replicated on the debug console.

5. Next, you can ask Mycroft a more advanced skill, such as looking up information on the internet. Ask, "Hey, Mycroft, how many ounces in a gallon?" Mycroft will use the internet to look up the answer and reply.

6. Next, you can change the wake word on the Mycroft website to something more appropriate – we did not name this robot Mycroft. We have been calling this robot Albert, but you can choose to call the robot anything you want. You may find that a very short name such as Bob is too quick to be a good wake word, so pick a name with at least two syllables. To do this, navigate to the Mycroft web page (`http://home.mycroft.ai`) and log in to your account, which we created back in *Step 4*. Click on your name in the upper right corner and select **Settings** from the menu. You can select several settings on this page, such as the type of voice you want, the

units of measurement, and time and date formats. Select **Advanced Settings**, which will take you to the page where we can change the wake word.

7. We change the first field, the **Wake word** field, to **Custom**. We change the next line to put in our custom wake word – "Hey, Albert."

8. We also need to look up the phonemes for this wake word. Go to *The CMU Pronouncing Dictionary* from Carnegie Mellon University (`http://www.speech.cs.cmu.edu/cgi-bin/cmudict`). Put in our phrase and you will get out the phoneme phrase. Copy and paste this phrase and go back to the Mycroft page to paste the phoneme phrase into the **Phonemes** field. You are done – don't change any of the other settings.

9. Hit **Save** at the top of the page before you navigate away.

You can test your new wake word back on the Jetson Nano. Start Mycroft up again in debug mode and wait for it to come up. Say your new wake phrase and enjoy the response. I have a standard test set of phrases to show Mycroft's skill at being the voice of our robot. Try the following:

*   Hey, Albert. What time is it?

*   Hey, Albert. What is the weather for tomorrow? Hey, Albert. How many ounces in a gallon?

*   Hey, Albert. Who is the king of England?

You should get the appropriate answers to these questions.

Mycroft has many other skills that we can take advantage of, such as setting a timer, setting an alarm, listening to music on Pandora, or playing the news. What we will be doing next is adding to these skills by creating our own that are specific to our room-cleaning robot.

Adding skills

The first skill we will create is a command to pick up toys. We are going to connect this command to the ROS to control the robot. Later on, we will add a skill to tell knock-knock jokes.

Designing our dialogs

Our first step is to design our dialog on how we will talk to the robot. Start by making a list of what ways you might tell the robot to pick up the toys in the playroom. Here is my list, which I generated using ChatGPT (version 3.5):

*   Hey robot, could you please start picking up all the toys?

*   It's time to tidy up. Can you gather all the toys for me?

*   Need your help, robot. Could you please pick up all the toys and put them in the toy bin?

- Let's clean up together. Begin by collecting all the toys, please.

- I need a clean room. Can you please start by picking up the toys scattered around?

- Dear robot, would you be so kind as to tidy up the room by gathering all the toys?

- Let's make this room spotless. Begin with collecting all the toys, please.

- It's clean-up time.

- Can you pick up all the toys?

- Pick up all the toys.

- You could collect all the toys and place them in the toy box.

- It's time to showcase your cleaning skills. Start by picking up all the toys, please.

- Could you please make the room tidy by picking up all the toys?

You will note that there are several keywords that are specific to the command to clean up the room. We have the word "clean," of course. We have the phrase "pick up," and "away." We also have the words "toys" or "toy", and finally, "mess." These keywords will cue the natural language processor and allow some variation in the exact words used.

Next, we write down what we want the robot to say back. We don't want the same canned response each time; it would be good to have some natural variation in the robot's responses. Here is my list of responses, created again by ChatGPT3, with a variety of robot attitudes represented:

- Command received – picking up toys.

- Of course, I will start cleaning the room right away.

- Understood, I will tidy up the room and ensure everything is in its proper place.

- Cleaning mode activated! I will make sure to leave your room spotless and organized.

- As you wish, I will take care of cleaning the room for you.

- No problem, I will use my cleaning capabilities to efficiently tidy up the room and remove any clutter.

- Consider it done! I will sweep, dust, and mop the room to make it pristine for you.

- Cleaning in progress! I will diligently clean up the room and ensure it's tidy and presentable.

- I'm on it! I will use my cleaning tools and techniques to make your room look spick and span.

- Acknowledged, I will clean up the room and ensure it's organized and ready for your use.

- Cleaning task accepted! I will work diligently to make your room clean and inviting, just the way you like it.

You can use as many of these as you want. The Mycroft system will grab a random phrase from this list. This gives us some room for creativity and gives the illusion that the robot is more intelligent than it really is. This type of response system enables us to quickly develop our dialogs.

In the next section, we'll look at how we can connect voice commands to the ROS interface.

Creating skills

In this section, we'll build the framework in Mycroft that allows it to associate our spoken phrases with the commands to be sent to the robot. This will involve adding each skill that we want the robot to possess.

Cleaning up the toys

First, let's add the most basic skill for Albert – cleaning the room by picking up toys. To do this, follow these steps:

1. Use the `mycroft-msk create` command, which helps us put together our skills in the proper format:

    ```
    mycroft-msk create
    ```

2. It will ask for a skill name. Let's use `cleanroomrobot-skill`.

3. Then it will ask for a class name and a repository name, for both of which I used `Cleanroomrobot`.

4. Enter a one-line description for your skill: `Pick up all of the toys in the room`.

5. Enter a long description, such as `Command the robot to detect toys, move to grab a toy, pick it up, and put it into the toybox`.

6. Enter some example phrases to trigger your skill:

    *   `Hey robot, could you please start picking up all the toys?`

    *   `It's time to tidy up. Can you gather all the toys for me?`

    *   `Can you pick up all the toys?`

7. Enter the following parameters:

    *   **Author**: `<your name here>`

    *   **Category**: `Productivity`

    *   **Secondary Category**: `IoT`

8. Entering tags makes it easier to search for your skill (although this is optional): `robot`, `cleanup`, `pick up`, and `toys`.

9. We will end up with a directory structure in /opt/Mycroft/skills/cleanroomrobot-skill like the following:

```
Cleanroomrobot-skill
    Git
    __pycache__
    Locale
        En-us
            cleanroomrobot.dialog
            cleanroomrobot.intent
    __init__.py
    LISCENSE.md
    Manifest.yml
    README.md
    Settingsmeta.yaml
```

10. Now we can populate the Python code that will activate the command to the robot. We need to edit the init.py file in the skill_pickup_toys directory that we copied from the template.

11. We are going to import the libraries we need from Mycroft (IntentBuilder, MycroftSkill, getLogger, and intent_handler). We also import rclpy, the ROS Python interface, and the ROS standard message String, which we use to send commands to the robot by publishing on the syscommand topic:

```
from mycroft import MycroftSkill, intent_handler, intent_file_
handler
import rclpy
from rclpy.node import Node
from std_msgs.msg import String, Int32MultiArray, Int32
from adapt.intent import IntentBuilder
from mycroft.util.log import getLogger
```

> **Note**
>
> MycroftSkill is a function that is called when one of its phrases is recognized by the Mycroft Intent Engine. As such, it has no body or main function, just a function definition for the create_skill function that instantiates a MycroftSkill object. The init function does most of the work of setting up the various dialogs, intent handlers, and vocabulary for the skill. This arrangement works very well in our limited environment of giving the robot commands or telling jokes.

12. The next line is the logger for Mycroft so that we can save our responses. Anything we put out to stdout, such as print statements, will end up in the log, or on the screen if you are in debug mode:

```
LOGGER = getLogger( name )
```

13. The next step is to set up a class for our skill object. We'll call it `Cleanroomrobot` to match what we defined previously:

```
class Cleanroomrobot(MycroftSkill):
    def __init__(self):
        MycroftSkill.__init__(self)

    def setRobotInterface(self,interfce):
        self.interface = interfce

    def initialize(self):
        pass  # just return for now
```

14. We set up the publisher for our `syscommand` topic in the ROS. This is how we send commands to the robot control program via the ROS publish/subscribe system. We will be publishing commands only, and the only message format we need is `String`:

```
pub = rospy.Publisher('/syscommand', String, queue_size=1000)
# define our service for publishing commands to the robot
control system # all our robot commands go out on the topic
syscommand
def pubMessage(str):
pub.publish(str)
```

15. Our Mycroft skill is created as a child object of the `MycroftSkill` object. We rename our skill object class to `CleanRoomSkill`:

```
class CleanRoomSkill(MycroftSkill):
def    init  (self):
super(CleanRoomSkill, self).  init  (name="PickupToys")
```

According to the template, Mycroft requires both an `init` method and an `initialize` method. These commands set up the intent in the Intent Builder part of Mycroft and register our handler when any of our phrases are spoken.

16. Next, we refer to the dialogs we built back in the *Creating skills* section with `require("CleanRoomKeyword")`, so be careful that all the spelling is correct:

```
def initialize(self):
        clean_room_intent = IntentBuilder("cleanroomrobot").
require("cleanroomrobot").build()
        self.register_intent(clean_room_intent, self.handle_
cleanroomrobot)
```

17. This next section creates our handler for when the system has recognized one of our phrases, and we want to perform the action for this command. This is where we kick off the publish command to the robot's control program via the ROS using the `pubMessage` function we defined earlier:

```
@intent_file_handler('cleanroomrobot.intent')
##@intent_handler('cleanroomrobot.intent')
def handle_cleanroomrobot(self, message):
    self.speak_dialog('cleanroomrobot')
    self.interface.cmdPublisher("CleanRoom")
```

18. We also need a `stop` function, where we can command the robot to stop cleaning, if necessary, to prevent any sort of *Mickey Mouse – Sorcerer's Apprentice* accident:

```
def stop(self):
    self.interface.cmdPublisher("STOPCleanRoom")
    pass
```

> **Note**
>
> In the movie *Fantasia*, Mickey Mouse acts out the part of the Sorcerer's Apprentice from a fairy tale. In the story, the Apprentice learns to animate a broom, which he teaches to fetch water from a well. The problem is the Apprentice never learned how to stop the enchantment, and soon the room is flooded.

19. We now need a block of code to create the skill in the program where we can associate the ROS interface to the robot into the skill. We will add a `create_skill` function to allow Mycroft to create the skill and to have a function pointer to enable the skill:

```
def create_skill():
    newSkill = Cleanroomrobot()
    newSkill.setRobotInterface(rosInterface())
    return newSkill
```

20. Next, we have the ROS interface. All we need to do is send a command to the robot to publish mode commands on our `RobotCmd` topic:

```
class rosInterface(Node):
    def __init__(self):
        super().__init__('mycroftROS') # node name
        self.cmdSubscribe = self.create_subscription(String,
'RobotCmd', self.cmdCallback,10)
        self.cmdPublisher = self.create_publisher(String,
'RobotCmd', 10)
```

```
def cmdCallback(self,msg):
    robotCmd = msg.data
```

We define our ROS interface and create a control node called `mycroftROS` to serve as our interface. Then we create a subscriber and publisher to the `RobotCmd` topic so we can send and receive commands from the ROS 2 interface.

21. The rest of the program is just housekeeping. We need to start up our ROS node, start the Mycroft logger, and instantiate the ROS interface object and the `cleanSkill` objects for ROS and Mycroft, respectively. Then we point the `cleanSkill` object to the ROS interface so they can communicate. Finally, we start the ROS 2 interface with the `.spin` function. When the program is stopped, we fall out of `.spin` and shut down our program:

```
## main ###
rclpy.init()
LOGGER = getLogger(__name__)
interface = rosInterface()
cleanSkill = Cleanroomrobot()
cleanSkill.setRobotInterface(interface)

rclpy.spin(interface)

rosInterface.destroy_node()
rclpy.shutdown()
```

22. In order for our skill to work, we need to copy our directory to `/opt/mycroft/skills`. From there, we can test it in debug mode. Remember that you have to source the ROS 2 directory (`source /opt/ros/foxy/local_setup.sh` and `source ~/ros2_ws/install/local_setup.sh`) or the program won't be able to find all of the inclusion files or ROS nodes.

Our next skill comes at the request of my grandson, William, who just adores knock-knock jokes. William is seven, so he is just the right age for this. Let's look at how we can implement this.

Telling jokes

In this section, we will handle the case where the robot is telling the knock-knock joke. As you probably know, knock-knock jokes are pun-based jokes that always take the same form:

Person 1: Knock, knock

Person 2: Who's there?

Person 1: Wooden

Person 2: Wooden Who?

Person 1: Wooden you like to know!

As you can see, the dialog is very simple. Several parts of it are standard, such as the first two lines – "Knock, knock" and "Who's there?" We can create a generic knock-knock joke in the following form:

1. Knock, knock.

2. Who's there?

3. `<word 1>`

4. `<word 1>` who?

5. `<punchline phrase>`

In defining our joke, you can see we just have two variable elements – the word in *Step 3*, and the punchline phrase in *Step 5*. Our word is repeated in *Step 4*.

We begin by creating a joke database of one-line jokes, which we will put in a text file. Since we just have two elements, we can separate them with a slash (/). Here is an example:

```
tarzan / tarzan stripes forever
orange / orange you glad I can tell jokes?
```

I'm providing you with a database of about 10 jokes in the files section of the repository for this chapter. Please feel free to add all of your favorites, or send them to me and I'll add them.

Now, let's look at the steps involved in telling the joke:

1. We will start, as with any skill, with the wake word, "Hey, Albert."

2. Then we need a phrase to indicate we want to hear a joke, so we will use variations of "Tell me a knock-knock joke," such as "I want to hear a knock-knock joke."

3. This will trigger our skill program to look up a joke. We will create several intents, or response capabilities, to respond to the user (or child) talking to the robot. We will start with the "Who's there?" dialog intent. That will let the robot know to proceed to the next part of the joke, which is to say our word.

4. Then we disable the "Who's there?" dialog and enable a dialog for listening for `<word>` and the phrase "who."

5. Then we can deliver the final part of the joke by reciting the punchline phrase, and we are done.

How can we implement this? You can follow these steps:

1. We start by creating our vocabulary files, of which we will need three. These are the things that the user will be saying to the robot. We have our first "tell me a knock, knock joke" phrase – so let's create a file called knockknock.voc (you can use any text editor to create the file) and put the following in it:

    ```
    Tell me a knock-knock joke Can I have a knock-knock joke Give me
    a knock-knock joke Play me a knock-knock joke
    ```

    Please note that the Mycroft STT system interprets the phrase "knock, knock" as knock-knock with a hyphen, so it is important to put that into our script.

2. Now our second vocabulary is just "Who's there," so we can create this as a second .voc file, whosthere.voc, which contains the line Whos there.

3. Our final line is a bit trickier. We really only care about the keyword "who" to trigger the punchline, so we can look only for that. Make a file called who.voc and put the one word who in it. Remember these all go in the dialog/en-us folder in our skill directory.

4. Now for our responses. We have one canned response, which is to reply to "tell me a knock-knock joke" with the phrase "knock, knock." We don't need any sophisticated dialog system, we just have the robot say the "knock, knock" phrase. To do this, we first import the libraries we need to call in this program, which are the MycroftSkill class and the intent_file_handler function:

    ```
    from mycroft import MycroftSkill, intent_file_handler
    ```

5. We define our skill as a child object of the MycroftSkill object – this is a standard object-oriented design. We are inheriting all of the functions and data of the MycroftSkill parent object and adding our own functionality to it. We create an initialize function and then call the init parent function to execute the code of the parent class as well. We are augmenting the functionality of the init parent function. Without this call, we would be replacing the init function with our own, and might have to duplicate a considerable amount of work:

    ```
    class Knockknock(MycroftSkill):
      def __init__(self):
        MycroftSkill.__init__(self)
    ```

6. The next step is to create our **intent handler**. The intent handler is called when the Intent Engine sees the keywords that indicate that this is what we want. Whenever the user asks, "Do you know any knock-knock jokes?" or phrases with that meaning, this code will be invoked. We put our phrases into the knockknock.intent file and place that file in the voc directory (which was dialog/voc-en):

    ```
    @intent_file_handler('knockknock.intent')
    def handle_knockknock(self, message):
    ```

We now need to pick a joke from our database of wonderful, witty knock-knock jokes. We will define our `pick_joke` function next:

```
name,punchline = self.pick_joke()
```

Here, we get two parts from the joke database:

- The name to say after "who's there"

- The punchline that ends the joke

7. We use the `get_response` function from `MycroftSkill` to have the robot make a statement and then wait for a reply, which will get turned into a text string and stored in the `response` variable:

```
response=self.get_response(announcement="knock, knock")
# response will always be "who's there"
response=self.get_response(announcement=name)
```

8. Now we are at the part where the robot says the name in response. For example, the user asks "who's there?" and the robot replies "Harold." What we are expecting next is for the user to say "Harold (or whatever name) who?" We will check our response, and see whether the word "who" is included. If it is not, we can prompt the user to follow along with the joke. We will only do this one time to keep from getting stuck in a loop. If they are not playing along, the robot will just continue:

```
# response will be "name who"
# if end of respose is not the word who, we can re-prompt
if "who" not in response:
  prompt = "You are supposed to say "+name+" who"
  response=self.get_response(announcement=prompt)
```

9. We have moved through the joke, so now we get to say the punchline, such as "Harold you like a hug?" (How would you like a hug?). The task is complete and we exit the routine; both the comedy routine and the program routine:

```
self.speak(punchline)
```

10. We need a function to read the joke database we defined earlier. As described earlier, the database has one knock-knock joke per line, with a forward slash (/) between the name and the punchline. We read all of the jokes, put them in a list, and then choose one at random using the (wait for it) `random.choice` function. We return the name and the punchline separately. We should only call this routine once per instance of the joke:

```
def pick_joke():
  jokeFile="knockknock.jokes"
  jfile = open(jokeFile,"r")
```

```
jokes = []
for jokeline in jfile:
   jokes.append(jokeline)
joke = choice(jokes)
jokeParts = joke.split("/")
name = jokeParts[0]
punchline = jokeParts[1]
return name, punchline
```

11. We finish the program by defining our instance of the Knockknock class and returning that object to the calling program, Mycroft:

```
def create_skill():
   return Knockknock()
```

Next, we'll discuss the other end of the knock-knock joke concept, which is to receive a joke – where the child wants to tell the robot a joke. If you know any seven-year-olds, then you know that this is a requirement also – the child will want to tell the robot a joke as well.

Receiving jokes

The receiving dialog is pretty simple as well. The user will say "knock, knock", which is the cue to the robot to go into the *receive knock-knock joke* mode. The robot then has only one response – "who's there." We could also add "who is there?" if we want to keep to the common sci-fi concept that robots do not use contractions.

> **Note**
>
> Data, the android from *Star Trek: The Next Generation*, stated many times he was not able to use contractions, although the writers slipped up from time to time.

In order to create our schema for our new Mycroft skill, we will be using the **Mycroft Skill Kit** (**MSK**). You can install MSK by typing pip3 install msk. MSK provides a dialog-driven approach to building skills that will make a framework, including all of the subdirectories for dialog files and vocabulary. This saves a lot of time, so let's try it out:

1. The following is the command for creating the *receive knock-knock joke* code:

```
$ msk create
Enter a short unique skill name (ie. "siren alarm" or "pizza
orderer"): receiveKnock
Class name: ReceiveKnockSkill
Repo name: receive-knock-skill
Looks good? (Y/n) y
Enter some example phrases to trigger your skill:
```

```
knock knock
-
Enter what your skill should say to respond:
who's there
Enter a one line description for your skill (ie. Orders fresh
pizzas from the store): This skill receives a knock knock joke
from the user
Enter a long description:
This is the other half of the Knock Knock joke continuum - we
are giving the robot the ability to receive knock knock jokes.
The user says knock knock, the robot responds whos there and so
on
>
Enter author: Francis Govers
Would you like to create a GitHub repo for it? (Y/n) Y
=== GitHub Credentials === Username: **********
Password:*********
Counting objects: 12, done.
Delta compression using up to 4 threads. Compressing objects:
100% (5/5), done.
Writing objects: 100% (12/12), 1.35 KiB | 0 bytes/s, done. Total
12 (delta 0), reused 0 (delta 0)
To https://github.com/FGovers/receive-knock-skill
* [new branch] master -> master
Branch master set up to track remote branch master from origin.
Created GitHub repo: https://github.com/FGovers/receive-knock-
skill Created skill at: /opt/mycroft/skills/receive-knock-skill
```

2. We can then either log in to the GitHub repository we just created (using your name rather than mine) and edit the program, or edit the source code at /opt/Mycroft/skills/receive-knock-skill. The program is still the init.py file.

3. We start with our imports, which are MycroftSkill and intent_file_handler. We will also need the time library to do some pauses:

```
from mycroft import MycroftSkill, intent_file_handler import
time
```

4. Here is our class definition for our ReceiveKnock class, which is a child class of the MycroftSkill object we imported. We start the init function by passing an init command back up to the parent class (MycroftSkill) and have it do its initialization. Then we add our custom functionality on top of that:

```
class ReceiveKnock(MycroftSkill):
  def __init__(self):
      MycroftSkill.__init__(self)
```

5. The next section is our intent handler for receiving a knock-knock joke. We use the @decorator to extend the intent handler, in this case, reading the parameters of the intent from a file called knock.receive.intent. The intent handler just has our two key words, the immortal phrase: knock, knock. We are fortunate that all jokes start exactly the same way, so we only have these two words.

    After the handle_knock_receive function has been activated by the Intent Engine seeing the phrase "knock, knock," we then get control passed to our handler. What is our next step? We reply with the single answer "Who is there?" You will remember we said robots do not use contractions. We use a different function to do this. We don't want to use another intent handler, but fortunately, Mycroft provides a free-form interface called get_response. You need to look up the documentation for this versatile function, but it makes our joke routine a lot simpler. The get_response function both lets us speak our reply and then receive whatever the user says next and store it as a string in the response variable:

    ```
    @intent_file_handler('knock.receive.intent')
    def handle_knock_receive(self, message):
      response =self.get_response('who.is.there')
    ```

    Now that we have our response, we can just repeat it back with the robot's voice, with the additional word "who?" So, if the child says, "Howard," the robot responds "Howard who?"

6. We use get_response again to have the robot speak and then record whatever the child or adult says next. We don't need it, but we want to have the robot's speech system listen to whatever is said next. We toss away the response, but insert our own comment to the joke from our dialog veryfunny.dialog, which is a file in the dialog directory. I created this file to hold responses to our jokes from the robot. I tried to make some responses that the grandchildren would find funny – I guess I can add "robot joke writer" to my resume, as I seem to have done this a lot in my career. After this, I added a sleep timer to allow everything to settle down before returning control. We include the standard stop function required of all MycroftSkills, and make our create_skill function make a ReceiveCall object and return it:

    ```
    response2= response + " who?"
    response3 =self.get_response(announcement=response2)
    self.speak_dialog('veryfunny')
    time.sleep(3)
    def stop(self):
      pass
    def create_skill():
      return ReceiveKnock()
    ```

You can get as creative as you want with the responses, but here are my suggestions:

- That was very funny!

- Ha ha ha.

- Very good joke.

- I like that one. Thank you!

- Ho HO! Ho.

- That was cute!

- I do not have a sound for a groan thththththpppppp!

Here is our directory structure and files for our receive knock-knock jokes skill:

```
receive-knock-skill directory:
init .py README.md
settingsmeta.json
./dialog/en-us:
knock.receive.dialog veryfunny.dialog
./vocab/en-us:
knock.receive.intent
```

Remember the local version of the skill goes in the `/opt/mycroft/skills/receive-knock-skill` directory. Now test to your heart's content – how many knock-knock jokes can you tell the robot?

Summary

This chapter introduced NLP for robotics and concentrated on developing a natural language interface for the robot that accomplished three tasks: starting the *pick up toys* process, telling knock-knock jokes, and listening to knock-knock jokes.

The concepts introduced included recognizing words by phonemes, turning phonemes into graphemes and graphemes into words, parsing intent from sentences, and executing computer programs with a voice interface. We introduced the open source AI engine, Mycroft, which is an AI-based voice assistant program that runs on the Jetson Nano. We also wrote a joke database to entertain small children with some very simple dialog.

In the next chapter, we'll be learning about **robot navigation** using landmarks, neural networks, obstacle avoidance, and machine learning.

Questions

1. Do some internet research on why the AI engine was named Mycroft. How many different stories did you find, and which one did you like?

2. In the discussion of intent, how would you design a neural network to predict command intent from natural language sentences?

3. Rewrite "Receive knock-knock jokes" to remember the jokes told to the robot by adding them to the joke database used by the "tell knock knock jokes" program. Is this machine learning?

4. Modify the "tell jokes" program to play sounds from a wave file, such as a music clip, as well as doing TTS.

5. The sentence structures used in this chapter are all based on English grammar. Other languages, such as French and Japanese, have different structures. How does that change the parsing of sentences? Would the program we wrote be able to understand Yoda?

6. Do you think that Mycroft's Intent Engine is actually understanding intent, or just pulling out keywords?

7. Describe the voice commands necessary to instruct the robot to drive to an object and pick it up without the robot being able to recognize the object. How many commands do you need?

8. From *Question 7*, work to minimize the number of commands. How many can you eliminate or combine?

9. Also from *Question 7*, how many unique keywords are involved? How many non-unique keywords?

Further reading

- *Python Natural Language Processing* by Jalaj Thanaki, Packt Publishing

- *Artificial Intelligence with Python* by Prateek Joshi, Packt Publishing

- Mycroft tutorial for developing skills is located at `https://mycroft.gitbook.io/mycroft-docs/developing_a_skill/introduction-developing-skills`

- Additional documentation for using Mycroft is located at `https://media.readthedocs.org/pdf/mycroft-core/stable/mycroft-core.pdf`

Part 3: Advanced Concepts – Navigation, Manipulation, Emotions, and More

In the last part of the book, we tackle more advanced topics including AI-based navigation and obstacle avoidance. We learn about decision trees and classification algorithms for unsupervised learning and then start an exciting chapter on creating a simulation of a robot personality. While we can't give a robot real emotions, we can create a simulation of emotion using state machines and Monte Carlo techniques. Finally, we end the book with a discussion of AI philosophy and a look at the future from the author's perspective, and provide advice for people wanting to pursue robotics and autonomy as a career.

This part has the following chapters:

7

Teaching the Robot to Navigate and Avoid Stairs

Let's have a quick review of our quest to create a robot that picks up toys. We've created a toy detector and trained the robot arm. What's next on our to-do list? We need to drive the robot to the location of the toy in order to pick it up. That sounds important.

This chapter covers **navigation** and **path planning** for our toy-grabbing robot helper. You have to admit that this is one of the most difficult problems in robotics. There are two parts to the task – figuring out where you are (localization), and then figuring out where you want to go (path planning). Most robots at this point would be using some sort of **simultaneous localization and mapping** (**SLAM**) algorithm that would first map the room, and then figure out where the robot is within it. But is this really necessary? First of all, SLAM generally requires some sort of 3D sensor, which we don't have, and a lot of processing, which we don't want to do. We can also add that it does not use machine learning, and this is a book about **artificial intelligence** (**AI**).

Is it possible to perform our task without making maps or ranging sensors? Can you think of any other robot that cleans rooms but does not do mapping? Of course you can. You probably even have a Roomba® (I have three), and most models do not do any mapping at all – they navigate by means of a pseudo-random statistical cleaning routine.

Our task in this chapter is to create a reliable navigation system for our robot that is adaptable to our mission of cleaning a single room or floor of toys, and that uses the sensors we already have.

The following topics will be covered in this chapter:

- Understanding the SLAM methodology
- Exploring alternative navigation techniques
- Introducing the Floor Finder algorithm for avoiding obstacles
- Implementing neural networks

Technical requirements

We require the **Robot Operating System Version 2** (**ROS 2**) for this chapter. This book uses the Foxy Fitzroy release: `http://wiki.ros.org/foxy/Installation`. This chapter assumes that you have completed *Chapter 6*, where we gave the robot a voice and the ability to receive voice commands. We will be using the Mycroft interface and voice text-to-speech system, which is called Mimic: `https://github.com/MycroftAI/mimic3`. You'll find the code for this chapter in the GitHub repository for this book at `https://github.com/PacktPublishing/Artificial-Intelligence-for-Robotics-2e`.

We will also be using the **Keras** library for Python (`https://keras.io`), which is a powerful library for machine learning applications and lets us build custom neural networks. You can install it using the following command:

```
pip install keras
```

You will also need **PyTorch**, which is installed with this command:

```
pip3 install torch torchvision torchaudio --index-url https://
download.pytorch.org/whl/cu118
```

Task analysis

As we do for each chapter, let's review what we are aiming to accomplish. We will be driving the robot around the house, looking for toys. Once we have a toy, we will take that toy to the toy box and put it away by dropping it into the toy box. Then, the robot will go look for more toys. Along the way, we need to avoid obstacles and hazards, which include a set of stairs going downward that would definitely damage the robot.

> **Note**
>
> I used a baby gate to cover the stairs for the first part of testing and put pillows on the stairs for the second part. There is no need to bounce the robot down the stairs while it is still learning.

We are going to start with the assumption that nothing in this task list requires the robot to know where it is. Is that true? We need to find the toy box – that is important. Can we find the toy box without knowing where it is? The answer is, of course, that the robot can just search for the toy box using its camera until it locates it. We developed a technique for recognizing the toy box back in *Chapter 4* with a neural network.

Now, if the robot was doing a bigger job, such as cleaning a 1,000,000-square-foot warehouse, then we would need a map. But our task is to clean a single 16 x 16 room. The time lost searching for the toy box is not all that significant, considering we can't get too far away, and we must drive to the toy box anyway. We will set this as a challenge, then, to accomplish our task without making a map.

> **Note**
>
> I once oversaw the evaluation of a robot system created at the Massachusetts Institute of Technology. They had a navigation system that did not use a map, and I was quite skeptical. In my defense, the robot actually got lost during the test. Now, I'm making a mapless navigator, and they are welcome to offer critique.

We also need to get the robot to do the following:

1. Navigate the room avoiding obstacles (toys and furniture) and hazards (stairs).
2. Find toys in the room (with the toy detector we created earlier).
3. Drive to a location where the robot arm can reach the toy.
4. Pick up the toy with the robot arm.
5. Carry the toy to the toy box.
6. Put the toy in the toy box.
7. Go and find another toy.
8. If there are no more toys, then stop.

We've covered finding the toy and picking it up in other chapters. In this chapter, we will discuss driving up to the toy to pick it up.

I'm a big fan of the movie *The Princess Bride*. It has sword fights, cliffs, two battles of wits, and **Rodents of Unusual Size** (**ROUS**). It also has a lesson in planning that we can emulate. When our heroes, Fezzik the Giant, Inigo Montoya, and Westley, plan on storming the castle to rescue the princess, the first things Westley asks are "What are our liabilities?" and "What are our assets?" Let's do this for our use case:

- **Our liabilities**: We have a small robot with a very limited sensor and compute capability. We have a room full of misplaced toys and a set of deadly stairs the robot can fall down.

- **Our assets**: We have a robot with omni wheels that can drive around, a voice, one camera, and a robot arm. The robot has a datalink via Wi-Fi to a control computer. We have this book. We have a toy box that is a distinctive color. And lots of **Toys of Usual Size** (**TOUS**).

The appropriate next step, whether we are designing robots or invading castles, is to do some brainstorming. How would you go about solving this problem?

We could use SLAM and make a map, then locate the robot on the map, and use that to navigate. Although we ultimately will not be following this method, let's quickly take a look at how it works.

Understanding the SLAM methodology

SLAM is a common methodology for navigating indoor robots. Before we get into the specifics, let's look at two key issues:

- The first problem we have in indoor robot driving is that we don't have a map

- The second problem we have is that we have no frame of reference to locate ourselves – GPS does not work indoors

That is two problems – we need a map, and then we need a way to locate ourselves on that map. While SLAM starts with the letter S for "simultaneous," in truth, most robots make a map, store it away, and then drive on it later. Of course, while maps are being made, the robot must make the map and then locate itself on the map – usually in the center.

How does SLAM work? The sensor usually associated with SLAM is the spinning LIDAR. You can think of LIDAR as laser radar – it uses a laser to measure the distance to objects and spins in a circle to collect data all around the robot.

We can summarize the SLAM method as follows:

1. The robot takes a measurement of the room by sweeping a laser rangefinder in a circle.

2. The data returned is a list of distance measurements, where the angular measure is a function of the position in the list. If we have a list of 360 measurements in a circle, then the first number in our list is 0 degrees, the next is 1 degree, and so on.

3. We can extract features in the LIDAR data by looking for corners, edges, jumps, and discontinuities.

4. We look at the angle and distance to each feature from succeeding measurements and create a function that gives the best estimate of how much the robot moved.

5. We use that information to transform the LIDAR data from the sensor-centric coordinate system to some sort of room coordinate system, usually by assuming that the starting position of the robot is coordinate 0,0. Our transform, or mathematical transformation, will be a combination of translation (movement) and rotation of the robot's body frame.

6. One way of estimating this transform is to use **particles**. We create samples of the robot's movement space at every point possible that the robot could have moved, and randomly place dots along all points. We compute the transform for each of these samples and then test to see which sample best fits the data collected. This is called a **particle filter** and is the technique I use for most of my SLAM projects.

For more details, you can refer to `https://www.cs.cmu.edu/~16831-f14/notes/` `F12/16831_lecture04_dfouhey.pdf`.

It can be difficult or impossible for SLAM to work in long, featureless hallways, for instance, as it simply has no information to work with – one lidar sweep looks just like the next. To help with this problem, many SLAM systems require the addition of other sensors to the robot, which measure wheel odometry or use optical flow to measure movement to provide additional data for the position estimate. The following is an illustration of a SLAM map of an office building made with ROS and displayed in RViz. The robot uses 500 particles for each LIDAR sample to estimate which changes in robot position best line up the lidar data with the data in the rest of the map. This is one of my earlier robot projects:

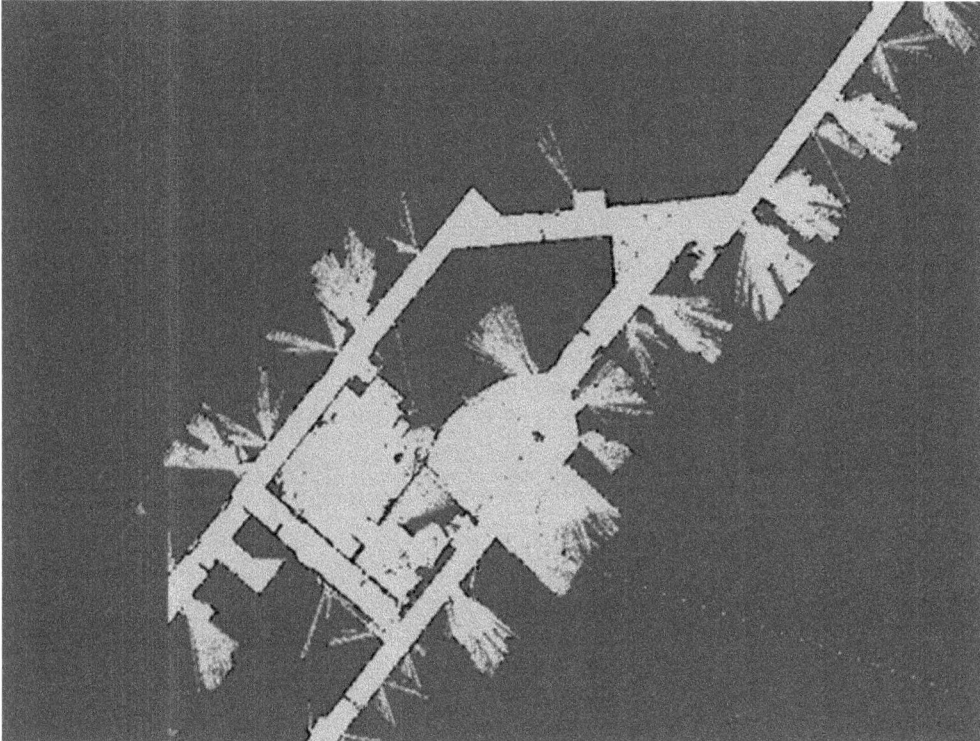

Figure 7.1 – A map generated by a SLAM navigation process

What we have to do in the SLAM process is as follows:

1. First, take a sweep that measures the distance from the robot to all of the objects in the room.
2. Then, we move the robot some distance – for example, three inches forward.
3. Then, we take another sweep and measure the distances again.
4. We now need to come up with a transformation that converts the data in the second sweep to line up with the data in the first sweep. To do this, there must be information in the two sweeps that can be correlated – corners, doorways, edges, and furniture.

You can get a very small robot LIDAR (e.g., the RPLidar from SLAMtec) for around $100, and use it to make maps. There is an excellent ROS package called *Hector Mapping* that makes using this LIDAR straightforward. You will find that SLAM is not a reliable process and will require several fits and starts to come up with a map that is usable. Once the map is created, you must keep it updated if anything in the room changes, such are re-arranging the furniture.

The SLAM process is actually very interesting, not for what happens in an individual scan, but in how scans are stitched together. There is an excellent video titled *Handheld Mapping in the Robocup 2011 Rescue Arena* that the authors of Hector SLAM, at the University of Darmstadt, Germany, put together, illustrating map making. It is available at the following link: `https://www.youtube.com/watch?v=F8pdObV_df4list=PL0E462904E5D35E29`.

I wanted to give you a quick heads-up on SLAM so that we could discuss why we are not going to use it. SLAM is an important topic and is widely used for navigation, but it is not the only way to solve our problem by any means. The weaknesses of SLAM for our purposes include the following:

- The need for some sort of sweeping sensor, such as LIDAR, ultrasound, or infrared, which can be expensive, mechanically complicated, and generate a lot of data. We want to keep our robot cheap, reliable, and simple.

- SLAM often works better if the robot has wheel odometers, which don't work on omni-wheeled vehicles such as our Albert. Omni wheels slide or skid over the surface in order to turn – we don't have Ackerman steering, such as a car with wheels that point. When the wheel skids, it is moving over the surface without turning, which invalidates any sort of wheel odometry, which assumes that the wheels are always turning in contact with a surface.

- SLAM does not deal with floorplans that are changing. The Albert robot has to deal with toys being distributed around the room, which would interfere with LIDAR and change the floorplan that SLAM uses to estimate position. The robot is also changing the floorplan as it picks up toys and puts them away.

- SLAM is computationally expensive. It requires the use of sensors to develop maps and then compares real-time sensor data to the map to localize the robot, which is a complex process.

- SLAM has problems if data is ambiguous, or if there are not enough features for the robot to estimate changes on. I've had problems with featureless hallways as well as rooms that are highly symmetrical.

So, why did I use this amount of space to talk about SLAM when I'm not going to teach you how to use it? Because you need to know what it is and how it works, because you may have a task that needs to make a map. There are lots of good tutorials on SLAM available, but very few on what I'm going to teach you next, which will be using AI to navigate safely without a map.

Exploring alternative navigation techniques

In this section, we'll look at some potential alternative methods for navigation that we could use for our robot now that we've ruled out the SLAM methodology:

- We could just drive around randomly, looking for toys. When we find a toy, the robot picks it up and then drives around randomly looking for the toy box. When it sees the toy box, it drives up to it and deposits the toy. But we still need a method to avoid running over obstacles. We could follow a process called **structure from motion** (**SfM**) to get depth information out of our single camera and use that to make a map. Structure from motion requires a lot of textures and edges, which houses may not have. It also leaves lots of voids (holes) that must be filled in the map. Structure from motion uses parallax in the video images to estimate the distance to the object in the camera's field of view. There has been a lot of interesting work in this area, and I have used it to create some promising results. The video image has to have a lot of detail in it so that the process can match points from one video image to the next. Here is a survey article on various approaches to SfM, if you are interested you can refer: `https://www.ijcit.com/archives/volume6/issue6/IJCIT060609.pdf`.

- You may have heard about a technique called **floor finding**, which is used in other robots and self-driving cars. I learned a great deal about floor finding from the sophisticated algorithm written by Stephen Gentner in the software package *RoboRealm*, which is an excellent tool for prototyping robot vision systems. You can find it at `http://www.roborealm.com`.

This floor finding technique is what we'll be using in this chapter. Let's discuss this in detail in the next section.

Introducing the Floor Finder technique

What I will be presenting in this chapter is my version of a Floor Finder technique that is different from RoboRealm, or other floor-finder algorithms, but that accomplishes the same results. Let's break this simple concept down for ease of understanding.

We know that the floor directly in front of the robot is free from obstacles. We use the video image pixels of the area just in front of the robot as an example and look for the same texture to be repeated farther away. We are matching the texture of the part of the image we know is the floor with pixels farther away. If the textures match, we mark that area green to show that it is drivable and free of obstacles. We will be using bits of this technique in this chapter. By the way, did you notice that I said *texture* and not *color*? We are not matching the color of the floor, because the floor is not all one color. I have a brown carpet in my upstairs game room, which still has considerable variation in coloring. Using color matching, which is simple, just won't cut it. We have to match the texture, which can be described in terms of color, intensity (brightness), hue, and roughness (a measure of how smooth the color of the surface is).

Let's try some quick experiments in this area with our image of the floor in my game room. There are several steps involved when doing this for real:

1. We start with the image we get from the camera. In order to accelerate processing and make the most efficient use of bandwidth, we set the native resolution of our camera – which has a full resolution of 1900 x 1200 – down to a mere 640 x 480. Since our robot is small, we are using a small computer – the Nvidia Jetson Nano.

2. We move that to our image processing program, using **OpenCV**, an open source computer vision library that also has been incorporated into ROS.

3. Our first step is to blur the image using the **Gaussian blur** function. The Gaussian blur uses a parabolic function to reduce the amount of high-frequency information in the image – it makes the image fuzzier by reducing the differences between neighboring pixels. To get enough blurring, I had to apply the blur function three times with a 5 x 5 **convolution kernel**. A convolution kernel is a matrix function – in this case, a 5 x 5 matrix of numbers. We use this function to modify a pixel based on its neighbors (the pixels around it). This smoothing makes the colors more uniform, reducing noise, and making the next steps easier. To blur the image, we take a bit from the surrounding pixels – two on either side – and add that to the center pixel. We discussed convolution kernels in *Chapter 4*.

4. We designate an area in front of the robot to be an area with a clear view of the floor. I used a triangular area, but a square area works as well. I picked each of the colors found within the triangle and grabbed all of the pixels that had a value with 15 units of that color. What does 15 units mean? Each color is encoded with an RGB value from 0 to 255. Our carpet color, brown, is around 162, 127, and 22 in red, green, and blue units. We select all the colors that are within 15 units of that color, which, for red, is from 147 to 177. This selects the areas of the image similar in color to our floor. Our wall is a very similar brown or beige, but fortunately, there is a white baseboard that we can isolate so that the robot does not try to climb the walls.

    Color is not the only way to match pixels on our floor. We can also look for pixels with a similar hue (shade of color, regardless of how bright or dark it is), pixels with the same **saturation** (darkness or lightness of color), and colors with the same value or **luminosity** (which is the same result as matching colors in a monochrome image or grayscale image). I compiled a chart illustrating this principle:

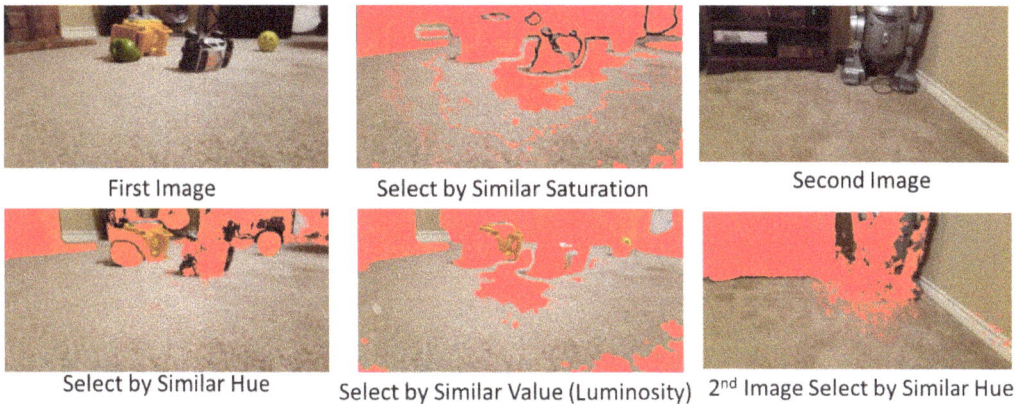

Figure 7.2 – Selecting pixels in an image by similarity of various attributes, such as color, hue, or saturation

The preceding figure shows the ability of various selection attributes (color, hue, saturation, and luminosity) as a tool to perform floor finding for our robot. The hue attribute seems to provide the best results in this test. I tested it on another image to be sure it was working. It seems not to separate out the baseboards, which are not part of the safe area to drive on.

5. We select all of the pixels that match our floor colors and paint them green – or, to be more correct, we create a mask region in a copy of the image that has all of the pixels we want to be designated somehow. We can use the number 10, for instance. We make a blank buffer the size of our image and turn all of the pixels in that buffer to 10, which would be the floor in the other image.

Performing an erode function on the masked data can help in this regard. There may be small holes or noise where one or two pixels did not match our carpet colors exactly – say there is a spot where someone dropped a cookie. The erode function reduces the level of detail in the mask by selecting a small region – for example, 3 x 3, and setting the mask pixel to 10 only if all of the surrounding pixels are also 10. This reduces the border of the mask by one pixel and removes any small speckles or dots that may be one or two pixels big. You can see from *Figure 7.3* that I was quite successful in isolating the floor area with a very solid mask. Given that we now know where the floor is, we paint the other pixels in our mask red, or some number signifying that it is unsafe to travel there. Let's use 255:

Floor Finder Concepts

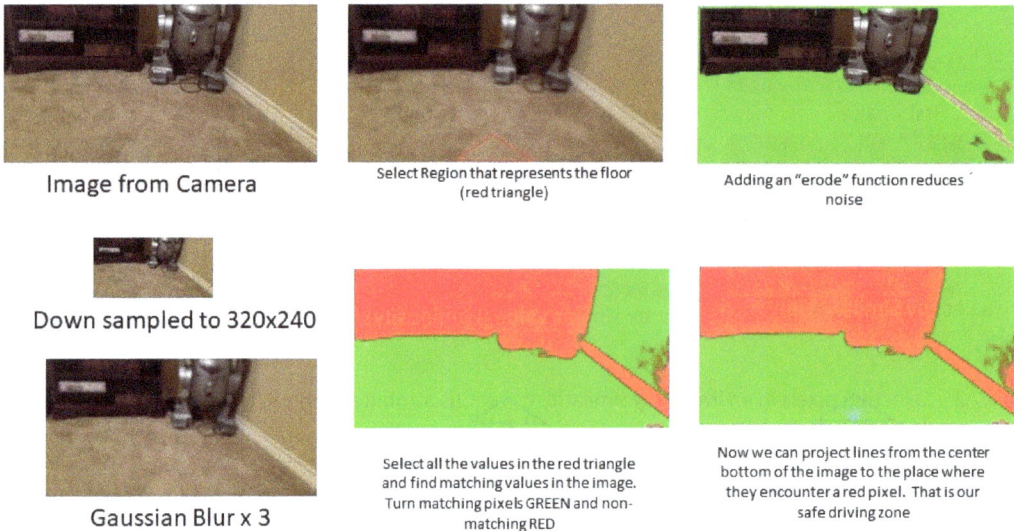

Image from Camera

Select Region that represents the floor
(red triangle)

Adding an "erode" function reduces
noise

Down sampled to 320x240

Select all the values in the red triangle
and find matching values in the image.
Turn matching pixels GREEN and non-
matching RED

Gaussian Blur x 3

Now we can project lines from the center
bottom of the image to the place where
they encounter a red pixel. That is our
safe driving zone

Figure 7.3 – My version of the floor finder algorithm

Note that it does a very good job in this case of identifying where it is safe to drive. The projected paths are required to prevent the robot from trying to drive up the wall. You get bonus points if you can identify the robot in the corner.

6. Our next step may take some thought on your part. We need to identify the areas that are safe to drive. There are two cases when using this process that may cause us problems:

- We may have an object in the middle of the floor by itself – such as a toy – that has green pixels on either side of it

- We may also have a concave region that the robot can get into but not out of

In *Figure 7.3*, you can see that the algorithm painted the wall pixels green since they match the color of the floor. There is a strong red band of no-go pixels where the baseboard is. To detect these two cases, we project lines from the robot's position up from the floor and identify the first red pixel we hit. That sets the boundary for where the robot can drive. You can get a similar result if you trace upward from the bottom of the image straight up until you hit a red pixel, and stop at the first one. Let's try the Floor Finder process again, but add some toys to the image so that we can be sure we are getting the result we want:

Floor Finder With Toys

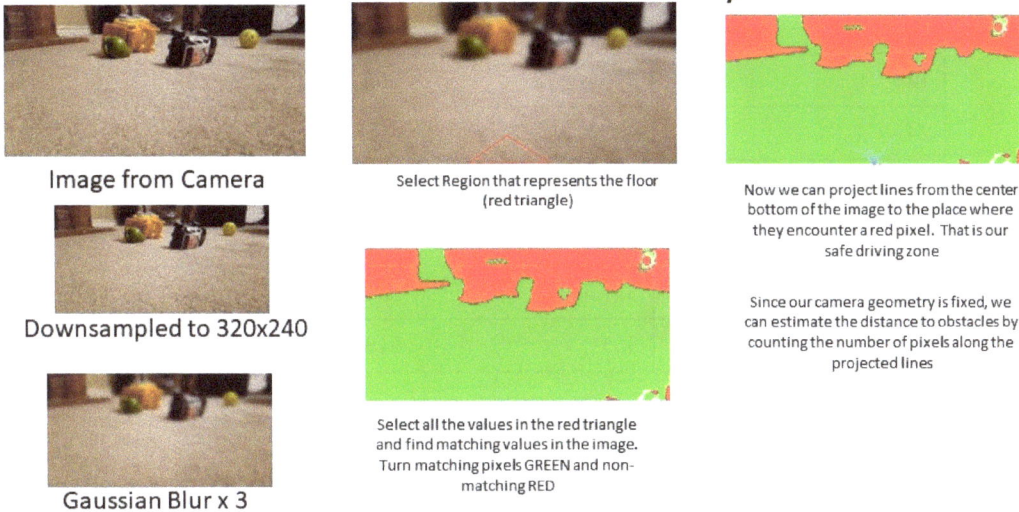

Image from Camera

Downsampled to 320x240

Gaussian Blur x 3

Select Region that represents the floor (red triangle)

Select all the values in the red triangle and find matching values in the image. Turn matching pixels GREEN and non-matching RED

Now we can project lines from the center bottom of the image to the place where they encounter a red pixel. That is our safe driving zone

Since our camera geometry is fixed, we can estimate the distance to obstacles by counting the number of pixels along the projected lines

Figure 7.4 – Adding toys to the image to determine if we are detecting toys as obstacles

That seems to be working well. We are able to find a good path to drive on. Keep in mind that we are constantly updating the obstacle view with the Floor Finder and updating our path as we drive. There is one shortcoming of this process. If a toy matches the color and texture of the carpet, then we might have a lot of difficulty finding it. You can add strip of masking tape to objects to deal with this issue, giving the camera something to see.

7. Another trick we can use with this process is to use the fixed camera geometry to do distance and size estimates. We have a "locked-down" camera – it is fixed in position on the robot, a set height from the floor, and, therefore, distance along the floor can be measured from the y value of the pixels. We would need to carefully calibrate the camera by using a tape measure and a box to match pixel values to the distance along the same path line we drew from the robot base to the obstacle. The distances will be nonlinear and only valid out to the distance the pixels continue to change. Since the camera is perpendicular to the floor, we get a certain amount of perspective effect that diminishes to 0 about 20 feet from the camera. My calibration resulted in the following table:

Measurement in inches	Distance from top	Distance from bottom
0	1080	0
12	715	365
24	627	453
36	598.3	481.7
48	581.5	498.5
60	571.8	508.2
72	565	515

Table 7.5 – Table of measurements comparing pixels to inches for scale

The following image shows the technique for measuring distance in the robot camera field of view. The object is located four feet away from the robot base along the tape measure. Albert uses a 180-degree fisheye lens on an HD-capable web camera. We need the wide field of view later in *Chapter 9* when we do navigation:

Figure 7.5 – Determining the scale of the pixels in our navigation camera image

One thing to watch out for is narrow passages that the robot will not fit into. We can estimate widths based on distance and pixels. One common robot technique is to put a border around all the obstacles equal to 1/2 the width of the robot. If there are obstacles on both sides, then the two borders will meet and the robot will know it does not fit.

In the next section, we will create a **convolutional neural network** (**CNN**) to take our images and turn them into robot commands – in essence, teaching our robot how to drive by seeing landmarks or features in the video image.

Implementing neural networks

So, what does a neural network do? We use a neural network to predict some association of an input with an output. When we use a CNN, we can associate a picture with some desired output. What we did in our previous chapter was to associate a class name (toys) with certain images. But what if we tried to associate something else with images?

How about this? We use a neural network to classify the images from our camera. We drive the robot around manually, using a joystick, and take a picture about four times a second. We record what the robot is doing in each picture – going forward, turning right, turning left, or backing up. We use that information to predict the robot's motion command given the image. We make a CNN, with the camera image as the input and four outputs – commands for go forward, go left, or go right. This has the advantage of avoiding fixed obstacles and hazards automatically. When we get to the stairs (remember that I have stairs going down in my game room that would damage the robot), the robot will know to turn around, because that is what we did in training – we won't deliberately drive the robot down the stairs during training (right?). We are teaching the robot to navigate the room by example.

You may be yelling at the book at this moment (and you should be) saying, "What about the toys?" Unless, of course, you are following my thought process and thinking to yourself, "Oh, that is why we just spent all that time talking about Floor Finder!" The neural network approach will get us around the room, and avoid the hazards and furniture, but will not help the robot to avoid toys, which are not in the training set. We can't put them in this training set because the toys are never in the same place twice. We will use the Floor Finder to help avoid the toys. How do we combine the two? The neural network provides the longer-range goal to the robot, and the Floor Finder modifies that goal to avoid local, short-range objects. In our program, we evaluate the neural network first and then use Floor Finder to pick a clear route.

On that theme, we are also going to pull another trick for training our robot. Since our floor surface is subject to change, and may be covered with toys, we will leave that part out of the training images. Before sending the image to the neural network, we'll cut the image in half and only use the top half. Since our camera is fixed and level with the floor, that gives us only the upper half of the room to use for navigation. Our image is a 180-degree wide angle, so we have a lot of information to work with. This should give us the resiliency to navigate under any conditions:

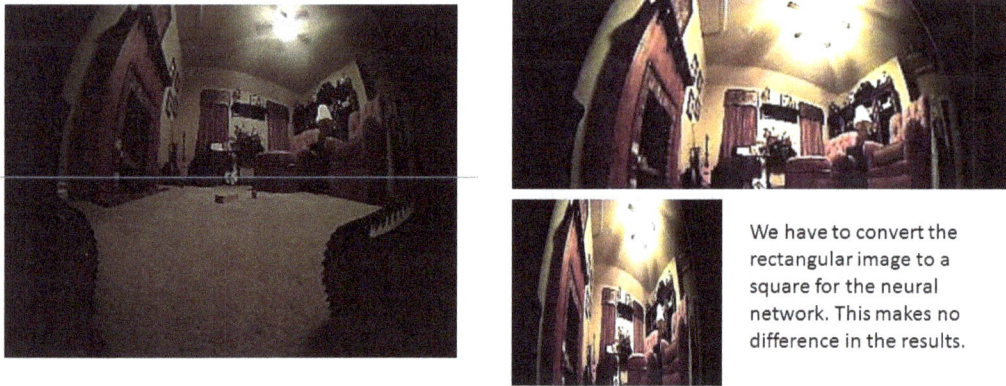

We have to convert the rectangular image to a square for the neural network. This makes no difference in the results.

Figure 7.6 – The training set for driving the robot only includes the top of the image

Our second problem is locating the toy box. For that, we need to create a new training set of images, which will represent an alternative driving pattern. We start the robot in various random locations, and then simply drive to the toy box. We use exactly the same process we used before for navigation – we are creating a training set that tells the robot how to get to the toy box. The trick is to get a good sample of every possible starting location. We do have a bit of a break – if a point on the map (a position in the room) is already on one path, we don't need to cover it again. In other words, all points that are included in another path don't need to be repeated. We still want to have at least 1,000 images to train from both sets of images – the one that explores the room, and the set that drives to the toy box.

I created a simple program that just lets the operator drive the robot with a joystick. It automatically takes a snapshot once a second. Each frame is labeled by simply looking at the value for the cmd_vel topic, which is how we control the motion of the robot base. If the angular velocity Z term (angular.z) is positive, we are turning right. If it is negative, we are turning left, and if the term is zero (you guessed it), we are driving straight ahead. I created an independent program that works with the camera and stores a snapshot whenever it receives a TAKE PIC LEFT, RIGHT, CENTER, or BACK command on the ROS syscommand topic. These programs will be in the GitHub repository for the book – I'm not going to include them here. We put each category of picture in its own subdirectory.

You can think of the neural network as working like this:

1. We present an image to the neural network.
2. It selects features from that image and then selects the images in the training database that are most like the features in the image provided.

3. Each picture in the training database is associated with a driving command (left, center, right). So, if the image most closely resembles an image where the robot turned left, then the network will return *turn left*.

Now let's look at these processes in greater detail.

Processing the image

Now, we have several steps to take before we can present our data to the neural network for training. Our camera on the robot has way too much resolution for what we need for the network, and we want to use the minimum amount of data in the neural network we can get away with:

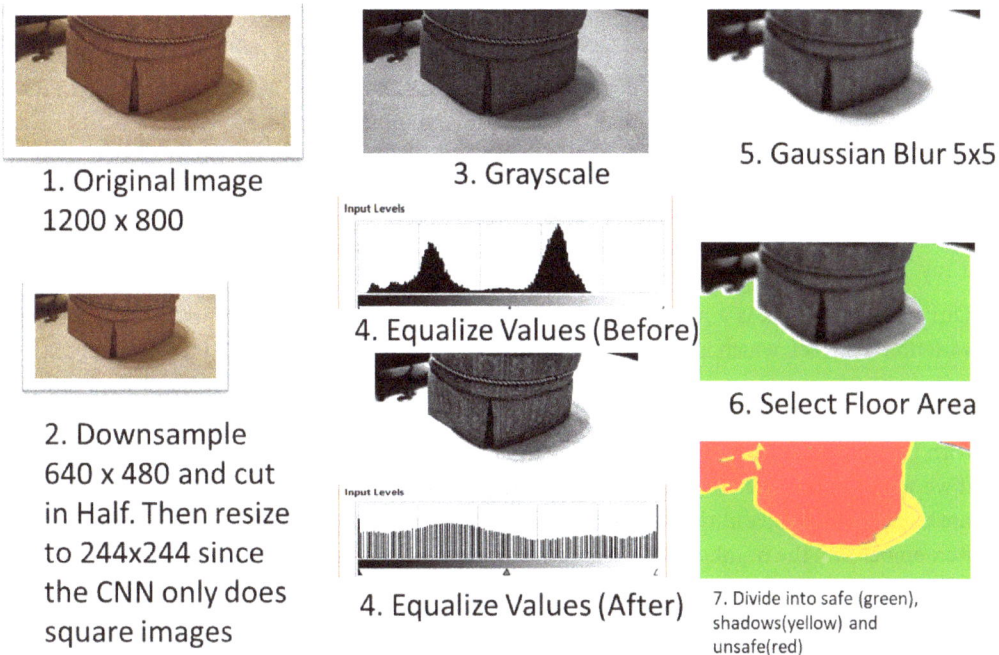

1. Original Image
1200 x 800

2. Downsample
640 x 480 and cut
in Half. Then resize
to 244x244 since
the CNN only does
square images

3. Grayscale

4. Equalize Values (Before)

4. Equalize Values (After)

5. Gaussian Blur 5x5

6. Select Floor Area

7. Divide into safe (green),
shadows(yellow) and
unsafe(red)

Figure 7.7 – Image processing for CNN

Let's break down this process to make this clearer:

1. The first image in the preceding figure represents our original image.

2. Our first step is to downsample the image to 640 x 480. We cut the image in half and keep only the top half, which eliminates the floor from our consideration. We resize the rectangular image to 244 x 244, which is an appropriate size for our neural network to process.

3. We convert the image to greyscale, so that we only have one channel to process, using this formula (proposed by the **National Television Standards Committee (NTSC)**):

$$Greyscale = 0.299 * R + 0.587 * G + 0.114 * B$$

4. Our next step is to equalize the image to take the entire range of possible values. The raw output of the camera contains neither pure white (255) nor pure black (0). The lowest value may be 53 and the highest, 180, for a range of 127. We scale the grayscale values by subtracting the low (53) and multiplying by the scale factor (127/255). This expands the range of the image to the full scale and eliminates a lot of the variation in lighting and illumination that may exist. We are trying to present consistent data to the neural network.

5. The next step is to perform a Gaussian blur function on the data. We want to reduce some of the high-frequency data in the image, to smooth out some of the edges. This is an optional step, and may not be necessary for your environment. I have a lot of detail in the robot's field of view, and I feel that the blur will give us better results. It also fills in some of the gaps in the grayscale histogram left by the equalization process in our previous step.

6. We have to normalize the data to reduce the scale from 0-255 to 0-1. This is to satisfy the artificial neural network's input requirements. To perform this operation, we just divide each pixel by 255. We also must convert the data from the OpenCV image format to a NumPy array. All of this is part of normal CNN preprocessing.

7. Our neural network is a nine-layer CNN. I used this common architecture because it is a variation of **LeNet**, which is widely used for this sort of task (http://vision.stanford. edu/cs598_spring07/papers/Lecun98.pdf). However, in our final step, rather than being a binary output determined by a binary classifier, we will use a **Softmax classifier** with four outputs – forward, left turn, or right turns. We can actually make more categories if we want to and have easy right and hard right turns rather than just one level of turns. I'm not using the full capability of the new omni wheels on my robot to keep this problem simple. Remember that the number of output categories must match our training set labels exactly.

In our CNN, the first six layers are pairs of CNNs with max pooling layers in between. This lets the network deal with incrementally larger details in the image. The final two layers are fully connected with **rectified linear units (ReLU)** activations. Remember that ReLU only takes the positive values from the other layers. Here is our final layer, which is a Softmax classifier with four outputs:

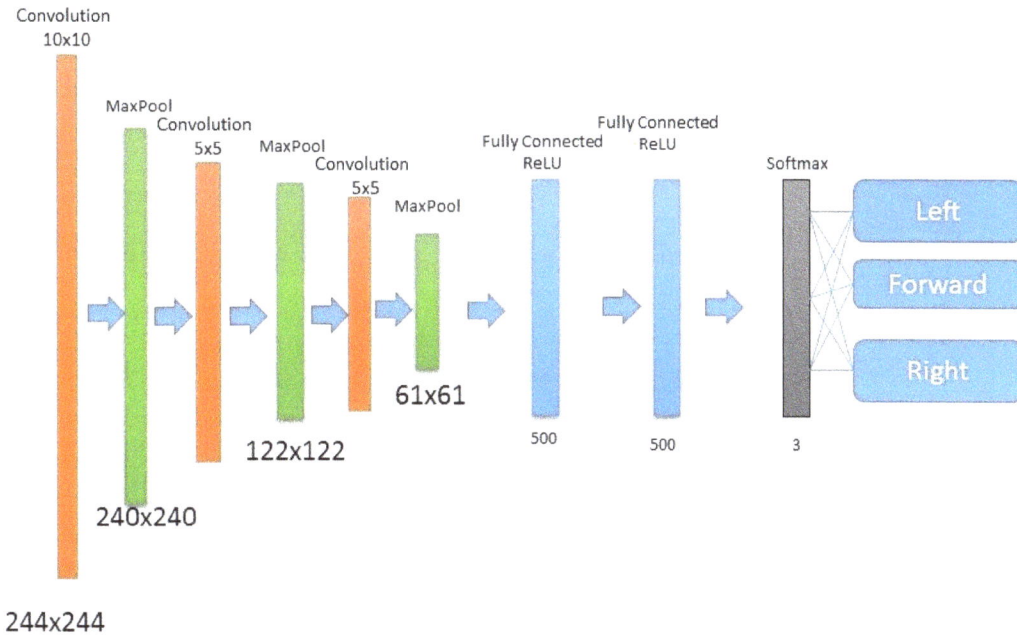

Figure 7.8 – Organization of our neural network

Like any other neural network training task, the next set of steps in the process involves splitting the input data into training sets and validation sets. Let's learn how to train the neural network next.

Training the neural network for navigation

We'll use 80% of our data on training and 20% on validation. We really can't use a process that sweetens the data by duplicating images with random rotations, as we did with the toy recognition program, since we are not just recognizing images, but using them for direction. Changing rotations would mess up our directions.

Now, let's put our training program together. This program was partially inspired by Adrian Rosebrock's *pyImageSearch* blog and by the paper *Deep Obstacle Avoidance* by Sullivan and Lawson at the Naval Research Lab. You can follow these steps:

1. We need to collect our training data by driving the robot around and recording our driving movements. This separated our data into three sets – left turn, right turn, and go straight. We have our training images in three subfolders to match our labels. We read in our data, associate it with the labels, and preprocess the data to present it to the neural network.

> **Note**
>
> I'm doing the training runs on my desktop computer, not on the Jetson Nano. We'll deploy on the Jetson Nano later with our fully trained networks.

2. Here are the imports that we need for this program – there are quite a few:

```
# import the necessary packages
from keras.preprocessing.image import ImageDataGenerator from
keras.optimizers import Adam
from sklearn.model_selection import train_test_split from keras.
preprocessing.image import img_to_array from keras.utils import
to_categorical
import matplotlib.pyplot as plt import numpy as np
import cv2 import os


from keras.models import Sequential
from keras.layers.convolutional import Conv2D
from keras.layers.convolutional import MaxPooling2D
from keras.layers.core import Activation
from keras.layers.core import Flatten
from keras.layers.core import Dense
from keras import backend as K
```

3. Here is the setup for the CNN:

-   We have three convolution layers, each followed by a `maxpooling` layer. Remember that each `maxpooling` layer will reduce the resolution of the image considered by the network by half, which is ¼ of the data, because we halve the width and the height. The convolution layers use the ReLU activation function since we don't want any negative pixel values.

-   After the convolution layers, we have two fully connected layers with 500 neurons each.

-   The final layer is our three neuron output layers, with a Softmax classifier that will output the percentage of each classification (left, right, and center). The output will look like (`0.8`, `0.15`, `0.05`), with three numbers that add up to 1.

This is a generic convolution network class that can be reused for other things, as it is a general multi-class image classification CNN:

```
class ConvNet():
    @staticmethod
    def create(width, height, depth, classes):
        # initialize the network
        network = Sequential()
        inputShape = (height, width, depth)
```

```
            # first set of CONV => RELU => POOL layers
            network.add(Conv2D(50, (10, 10), padding="same", input_
    shape=inputShape))
            network.add(Activation("relu"))
            network.add(MaxPooling2D(pool_size=(2, 2), strides=(2,
    2)))

            # second set of CONV => RELU => POOL layers
            network.add(Conv2D(50, (5, 5), padding="same"))
            network.add(Activation("relu"))
            network.add(MaxPooling2D(pool_size=(2, 2), strides=(2,
    2)))

            # third set of CONV => RELU => POOL layers
            network.add(Conv2D(50, (5, 5), padding="same"))
            network.add(Activation("relu"))
            network.add(MaxPooling2D(pool_size=(2, 2), strides=(2,
    2)))

            # Fully connected ReLU layers
            network.add(Flatten())
            network.add(Dense(500))
            network.add(Activation("relu"))
            network.add(Dense(500))
            network.add(Activation("relu"))

            # softmax classifier
            network.add(Dense(classes))
            network.add(Activation("softmax"))

            # return the constructed network architecture
            return network
```

4. Now, we set up our learning regime. We will run 25 training runs, with a learning rate of 0.001. We set a batch size of 32 images per batch, which we can reduce if we end up running out of memory:

```
EPOCHS = 25 LEARN_RATE = 1e-3
BATCH = 32 # batch size - modify if you run out of memory
```

5. The next section loads all of our images. We set the path here where the images reside. We put the three types of training images in folders named left, right, and center:

```
print ("Loading Images")
images=[]
labels=[]
#location of your images
imgPath = "c:\users\fxgovers\documents\book\chapter7\train\"
imageDirs=["left","right","center"]

for imgDir in imageDirs:
 fullPath = imgPath + imgDir
 # find all the images in this directory
 allFileNames =
 os.listdir(fullPath) ifiles=[]
 label = imgDirs.index(imgDir) # use the integer version of the
 label # 0= left, 1 = right, 2 = center
 for fname in allFileNames:
   if ".jpg" in fname:
      ifiles.append(fname)
```

6. Now, you can refer back to my diagram (*Figure 7.7*) of the process we will go through to preprocess the images. We will cut the image in half and just process the upper half of the picture. Then, we reduce the image to 244 x 244 to fit into the neural network, which needs square images. We will convert the image to grayscale (black and white) since we don't need to consider color, just shapes. This cuts our data down further. We will equalize the image, which rescales the range of gray colors to fill the whole area from 0 to 255. This evens out the illumination and sets the contrast:

```
# process all of the images
for ifname in ifiles:
   # load the image, pre-process it, and store it in the data
list image = cv2.imread(ifname)
   # let's get the image to a known size regardless of what was
collected
   image = cv2.resize(image, (800, 600))
   halfImage = 800*300 # half the pixels
   # cut the image in half -we take the top half
   image = image[0:halfimage]
   #size the image to what we want to put into the neural network
image=cv2.resize(image, (224,224))
   # convert to grayscale
   image = cv2.cvtColor(image, cv2.COLOR_BGR2GRAY) #equalize the
image to use the full range from 0 to 255 # this gets rid of a
lot of illumination variation
   image = cv2.equalizeHist(image)
```

7. Next, we have the Gaussian blur. This is an optional item – you may want to remove it if your room does not have a lot of detail. My game room has lots of furniture, so I think reducing the noise will improve performance:

```
# gaussian blur the image to remove high frequency noise # we
use a 5x kernel
image = cv2.GaussianBlur(img,(5,5),0)
```

8. We convert the image to a NumPy array of floats scaled from 0 to 1, instead of a set of integers from 0 to 255. This neural network toolkit only permits NumPy arrays for inputs. We also put the number associated with the labels (left = 0, right=1, and center = 2) into the matching labels NumPy array:

```
# convert to a numpy array image = img_to_array(image)
# normalize the data to be from 0 to 1
image2 = np.array(image, dtype="float") / 255.0 images=images.
append(image) labels.append(label)
labels = np.array(labels) # convert to array
```

9. We split the data into two parts – a training set that we use to train the neural network, and the testing set that we validate the training set with. We'll use 80% of the image samples for training and 20% for testing:

```
# split data into testing data and training data 80/20
(trainData, testData, trainLabel, testLabel) = train_test_
split(data, labels, test_size=0.20, random_state=42)
```

10. We have to convert the labels to a tensor, which is just a particular data format:

```
# convert the labels from integers to vectors
trainLabel = to_categorical(trainLabel, num_classes=3) testLabel
= to_categorical(testLabel, num_classes=3)
```

11. Now, we build our actual neural network by instantiating the ConvNet object, which actually builds our CNN in Keras. We set up the optimizer, which is **Adaptive Moment Estimation (ADAM)**, a type of adaptive gradient descent. ADAM acts against the error gradient like a heavy ball with friction – it has some momentum, but does not pick up speed quickly:

```
# initialize the artificial neural network print("compiling
CNN...")
cnn = ConvNet.build(width=224, height=224, depth=1, classes=3)
opt = Adam(lr=LEARN_RATE, decay=LEARN_RATE / EPOCHS) model.
compile(loss="categorical_crossentropy", optimizer=opt,
metrics=["accuracy"])
```

12. We train the network in this step. This will take quite some time to complete – from 15 minutes to an hour or two – depending on how many images you have. We want the training to come out somewhere above 80%. If not, add some epochs to see where the learning curve levels off.

If that still does not do the trick, you need more training images. I'm aiming for 1,000 images in each set, which is roughly 50 minutes of driving the robot around:

```
# train the network
print("Training network. This will take a while")
trainedNetwork = model.fit_generator(aug.flow(trainImage,
trainLabel, batch_size=BATCH),
validation_data=(testImage, testLable), steps_per_
epoch=len(trainImage) // BATCH,
epochs=EPOCHS, verbose=1) # save the model to disk
print("Writing network to disk") cnn.save("nav_model")
```

13. We are all done now, so we save the model we created to disk so that we can transfer it to the robot's computer, the Nvidia Jetson Nano.

14. Now, make your second training set of driving from random locations to the toy box. Pick random spots and use the joystick to drive the robot to the toy box from each. Keep going until you have 1,000 images or so. Run these through the training program and label this model toybox_model by changing the last line of the program:

```
cnn.save("toybox_model")
```

This is great – we have built and trained our neural network. Now, we need to put it to use to drive the robot around, which we'll do in the next section.

CNN robot control implementation

We need to combine a program that sends out ROS commands with our neural network classification process. I added some commands through the ROS syscommand topic, which I use for non-periodic commands to my robots. syscommand just publishes a string, so you can use it for just about anything. You can follow these steps:

1. We start with our imports from ROS, OpenCV2, and Keras, as we will be combining functions from all three libraries:

```
import roslib import sys import rospy import cv2
from std_msgs.msg import String
from sensor_msgs.msg import Image
from geometry_msgs.msg import Twist
from cv_bridge import CvBridge, CvBridgeError
from keras.preprocessing.image import img_to_array
from keras.models import load_model
import numpy as np
```

2. This first section is the ROS interface. I like to encapsulate the ROS interface this way, with all of the publish and subscribe in one place. We have several topics to set up – we need to be able to send and receive commands on the `syscommand` topic. We will be publishing commands to the robot's motors on the `cmd_vel` topic. We receive images from the camera on `image_topic`. We use callbacks to handle the event when a topic is published elsewhere on the robot. These can be called at any time. We have more control when we publish to a topic, which is handled using the `pubTwist` and `pubCmd` methods. I added flags to received commands and images so that we don't accidentally process the same image or command twice:

```
class ROSIF():
 def  init (self):
   self.bridge = CvBridge()
   self.image_sub = rospy.Subscriber("image_topic",Image,self.
callback)
   self.cmd_sub = rospy.Subscriber( "syscommand",String,self.
cmdCallback) self.cmd_pub = rospy.Publisher(
"syscommand",String,queue_size=10)
   self.twist_pub = rospy.Publisher("cmd_vel",Twist,queue_
size=10)
   self.newImage = False
   self.cmdReceived=""

 def callback(self):
  try:
   self.image = self.bridge.imgmsg_to_cv2(data, "bgr8")
   self.newImage = True
  except CvBridgeError as e:
   print(e)
 def cmdCallback(self,data):
  # receieve a message on syscommand
  self.cmdReceived = data.data

 def getCmd(self):
  cmd = self.cmdReceived
  self.cmdReceived = "" # clear the command so we dont do it
twice
  return cmd
```

3. This next function is the means for the rest of the program to get the latest image from the camera system, which is published on ROS on `image_topic`. We grab the latest image and set the `newImage` variable to `False`, so that we know next time whether we are trying to process the same image twice in a row. Each time we get a new image, we set `newImage` to `True`, and each time we use an image, we set `newImage` to `False`:

```
def getImage(self):
  if self.newImage=True:
    self.newImage = False
    # reset the flag so we don't process twice return self.image

    self.newImage = False
    # we send back a list with zero elements
    img = []
    return img
```

4. This section sends speed commands to the robot to match what the CNN output predicts for us to do. The output of the CNN is one of three values: left, right, or straight ahead. These come out of the neural network as one of three enumerated values – 0, 1, or 2. We convert them back to left, right, and center values, and then use that information to send a motion command to the robot. The robot uses the `Twist` message to send motor commands. The `Twist` data message is designed to accommodate very complex robots, quadcopters, and omni-wheel drive systems that can move in any direction, so it has a lot of extra values. We send a `Twist.linear.x` command to set the speed of the robot forward and backward, and a `Twist.angular.z` value to set the rotation, or turning, of the base. In our case, a positive `angular.z` rotation value goes to the right, and a negative value to the left. Our last statement publishes the data values on the `cmd_vel` topic as a `Twist` message:

```
# publishing commands back to the robot
def pubCmd(self,cmdstr):
  self.cmd_pub.publish(String(cmdstr)):

def pubTwist(self,cmd):
  if cmd == 0: # turn left
    turn = -2
    speed = 1
  if cmd==1:
    turn = 2
    speed = 1
  if cmd ==3:
    turn=0
    speed = 1 # all stop
  if cmd==4:
    turn = 0
```

```
    speed = 0
cmdTwist = Twist()
cmdTwist.linear.x = speed
cmdTwist.angular.z = turn self.twist_pub.publish(cmdTwist)
```

5. We create a function to do all of our image processing with one command. This is the exact replica of how we preprocessed the images for the training program – just as you might think. You may think it a bit strange that I scale the image up, only to then scale it down again. The reason for this is to have detail for the vertical part of the image. If I scaled it down to 240 x 240 and then cut it in half, I would be stretching pixels afterward to get it square again. I like having extra pixels when scaling down. The big advantage of this technique is that it does not matter what resolution the incoming image is at – we will end up with the correctly sized and cropped image.

The other steps involve converting the image to grayscale, performing an equalization on the contrast range, which expands our color values to fill the available space, and performing a Gaussian blur to reduce noise. We normalize the image for the neural network by converting our integer 0-255 grayscale values to floating point values from 0 to 1:

```
def processImage(img):
# need to process the image
image = cv2.resize(image, (640, 480))
halfImage = 640*240 # half the pixels
# cut the image in half -we take the top half image =
image[0:halfimage]
#size the image to what we want to put into the neural network
image=cv2.resize(image, (224,224))
# convert to grayscale
image = cv2.cvtColor(image, cv2.COLOR_BGR2GRAY)
 #equalize the image to use the ful
    image = cv2.equalizeHist(image)
# gaussian blur the image to remove high freqency noise # we use
a 5x kernel
image = cv2.GaussianBlur(img,(5,5),0) # convert to a numpy array
image = img_to_array(image)
# normalize the data to be from 0 to 1
image2 = np.array(image, dtype="float") / 255.0 return image2
```

6. Now that we're set up, we go into the main program. We have to initialize our ROS node so that we can talk to the ROS publish/subscribe system. We create a variable, mode, that we use to control what branch of processing to go down. We make an interface to allow the operator to turn the navigation function on and off, and to select between normal navigation and our toy-box-seeking mode.

In this first section, we will load both neural network models that we trained before:

```
# MAIN PROGRAM
ic = image_converter()
rosif = ROSIF()
rospy.init_node('ROS_cnn_nav')
mode = "OFF"
# load the model for regular navigation
navModel = load_model("nav_model")
toyboxModel = load_model("toybox_model")
```

7. This section begins the processing loop that runs while the program is active. Running `rospy.spin()` tells the ROS system to process any message that may be waiting for us. Our final step is to pause the program for 0.02 seconds to allow the Raspberry Pi to process other data and run other programs:

```
while not rospy.is_shutdown():
    rospy.spin()
    time.sleep(0.02)
```

So, that concludes our navigation chapter. We've covered both obstacle avoidance and room navigation using a neural network to teach the robot to drive about using landmarks on the ceiling – and without a map.

Summary

This chapter introduced some concepts for robot navigation in an unstructured environment, which is to say, in the real world, where the designers of the robot don't have control over the content of the space. We started by introducing SLAM, along with some of the strengths and weaknesses of map-based navigation. We talked about how Roomba navigates, by random interaction and statistical models. The method selected for our toy-gathering robot project, Albert, combined two algorithms that both relied mostly on vision sensors.

The first was the Floor Finder, a technique I learned when it was used by the winning entry in the DARPA Grand Challenge. The Floor Finder algorithm uses the near vision (next to the robot) to teach the far vision (away from the robot) what the texture of the floor is. We can then divide the room into things that are safe to drive on, and things that are not safe. This deals with our obstacle avoidance. Our navigation technique used a trained neural network to identify the path around the room by associating images of the room from the horizon up (the top half of the room) with directions to travel. This also served to teach the robot to stay away from the stairs. We discarded the bottom half of the room from the image for the neural network because that is where the toys are. We used the same process to train another neural network to find the toy box.

This process was the same as we saw in *Chapter 4*, but the training images were all labeled with the path from that spot to the toy box. This combination gave us the ability to teach the robot to find its way around by vision, and without a map, just like you do.

In the next chapter, we'll cover classifying objects, and review some other path-planning methods.

Questions

1. Regarding SLAM, what sensor is most commonly used to create the data that SLAM needs to make a map?

2. Why does SLAM work better with wheel odometer data available?

3. In the Floor Finder algorithm, what does the Gaussian blur function do to improve the results?

4. The final step in the Floor Finder is to trace upward from the robot position to the first red pixel. In what other way can this step be accomplished (referring to *Figure 7.3*)?

5. Why did we cut the image in half horizontally before doing our neural network processing?

6. What advantages does using the neural network approach provide that a technique such as SLAM does not?

7. If we used just a random driving function instead of the neural network, what new program or function would we have to add to the robot to achieve the same results?

8. How did we end up avoiding the stairs in the approach presented in the chapter? Do you feel this is adequate? Would you suggest any other means for accomplishing this task?

Further reading

- *Deep Obstacle Avoidance* by Sullivan and Lawson, published by Naval Research Labs, Rosebrock, Adrian.

- *Artificial Intelligence with Python Cookbook* by Ben Auffarth, Packt Publishing, 2020

- *Artificial Intelligence with Python – Second Edition*, by Prateek Joshi, Packt Publishing, 2020

- *Python Image Processing Cookbook* by Sandipan Dey, Packt Publishing, 2020

8
Putting Things Away

Imagine that you have to get to Grandma's house, which, according to legend, is *over the hills and through the woods*, and two states away. That would be two countries away if you live in Europe. To plan your trip, you can start in one of two ways. Ignoring the fact that Google has taken away most map reading and navigation skills from today's youth, you would get out a map and do one of the following:

- Start at your house and try to find the roads that are closest to a straight line to Grandma's house
- Start at Grandma's house and try to find roads leading to your home

From either direction, you will find that the road or path you seek forks, intersects, changes, meanders, and may even come to a dead end. Also, all roads are not created equally – some are bigger, with higher speed limits, and some are smaller, with more stop signs. In the end, you pick your route by the combination of decisions that results in the lowest cost. This cost may be in terms of *time* – how long to get there. It may be in terms of *distance* – how many miles to cover. Or it may be in *monetary* terms – there is a toll road that charges an extra fee.

In this chapter, we will be discussing several ways to solve problems involving choosing a chain of multiple decisions where there is some metric – such as cost – to help us select which combination is somehow the best. There is a lot of information here that is widely used in robotics, and we will be expanding our horizons a bit beyond our toy-grabbing robot to look at robot path planning and decision-making in general. These are critical skills for any robotics practitioner, so they are included here. This chapter covers the basics of decision-making processes for **artificial intelligence** (**AI**) where the problem can be described in terms of either a **classification problem** (determining whether this situation belongs to one or more groups of similar situations) or a **regression problem** (fitting or approximating a function that can be a curve or a path). Finally, we will be applying two approaches to our robot problem – an expert system and random forests.

This chapter will cover the following topics:

- Decision trees and random forests
- Path planning, grid searches, and the A* (A-star) algorithm

- Dynamic planning with the D* (D-star) technique
- Expert systems and knowledge bases

At first glance, the concepts we will cover in this section – namely, path planning, decision trees, random forests, grid searches, and GPS route finders – don't have much in common, other than all being part of computer algorithms used in AI. From my point of view, they are all basically the same concept and approach problems in the same way.

Technical requirements

The one tool we use for this chapter, you should have already installed from earlier chapters – **scikit-learn** (`http://scikit-learn.org/stable/developers/advanced_installation.html`).

Or, if you have the `pip` installer in Python, you can install it using the following command:

```
pip install -U scikit-learn
```

You'll find the code for this chapter at `https://github.com/PacktPublishing/Artificial-Intelligence-for-Robotics-2e`.

Task analysis

Our task in this chapter is one that you may have been waiting for if you have been keeping score since *Chapter 3*, where we discussed our storyboards. We need to navigate around the room on our wheels and find a path to our destination, whether that is picking up a toy or driving to a toybox.

To achieve this, we will be using **decision trees**, **classification** (a type of **unsupervised learning**), **fishbone diagrams**, which are good for troubleshooting, and finally, **path planning**.

Introducing decision trees

The concept of a **decision tree** is fairly simple. You are walking down the sidewalk and come to a corner. Here, you can go right, turn left, or go straight ahead. That is your decision. After making the decision – to turn left – you now have different decisions ahead of you than if you turned right. Each decision creates paths that lead to other decisions.

As we are walking down the sidewalk, we have a goal in mind. We are not just wandering around aimlessly; we are trying to get to some goal. One or more combinations of decisions will get us to the goal. Let's say the goal is to get to the grocery store to buy bread. There may be four or five paths down sidewalks that will get you to the store, but each path may be different in length or may have different paths. If one path goes up a hill, that may be harder than taking the level path. Another path may have you wait at a traffic light, which costs time. We assign a value to each of these attributes and generally want to pick the path with the lowest cost, or the highest reward, depending on the problem.

In the following decision tree, we can break down the actions of the robot in order to pick up a toy. We start by looking at the toy aspect ratio (the length versus width of the bounding box we detected in *Chapter 4*). We adjust the wrist of the robot arm based on the narrowest part of the toy. Then, we try to pick up the toy with that wrist position. If we are successful, we lift the toy off the ground and carry it to the toybox. If we fail, we try another position. After trying all of the positions, we go on to the next toy and try to come back to this toy later, hopefully from a different angle. You can see the utility of breaking down our actions this way, and it ends up that decision trees are useful for a lot of things, as we will see in this chapter:

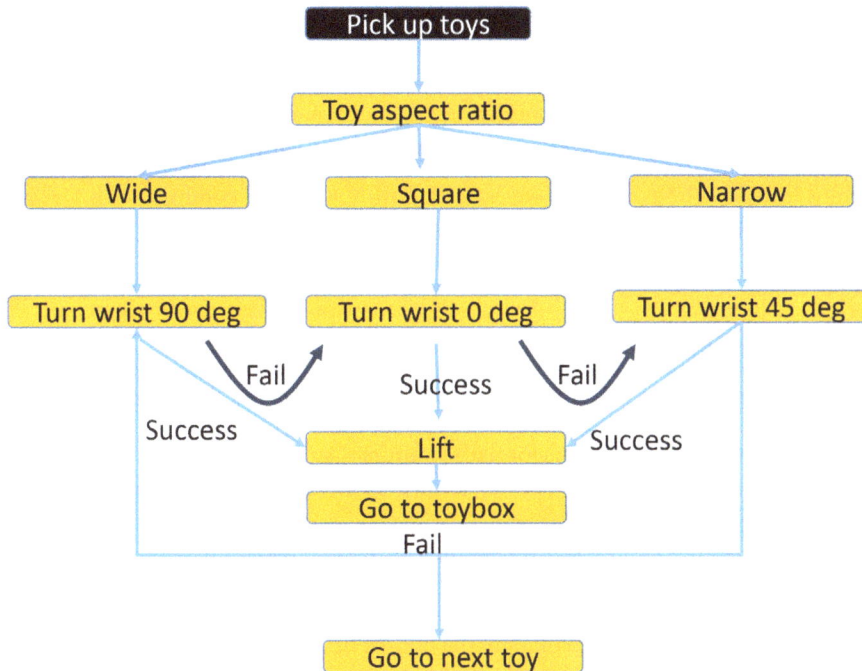

Figure 8.1 – A simple decision tree on how to pick up toys

The general problem with decision tree-type problems is one of *exponential growth*. Let's consider a chess game, a favorite problem set for AI. We have 20 choices for an opening move (8 pawns and 2 knights, each with 2 possible moves). Each of these 20 moves has 20 possible next moves, and so on. So the first move has 20 choices, and the second move has 400 choices. The third move has 197,281 choices! We soon have a very, very large decision tree as we try to plan ahead. We can say that each of these possible decisions is a **branch**, the state we are in after making the decision is a **leaf**, and the entire conceptual structure is a decision tree.

> **Note**
> The secret to working with decision trees is to ruthlessly prune the branches so you consider as few decisions as possible.

There are two ways to deal with a decision tree (actually, there are three – see if you can guess the third before I explain it):

- The first way is to start at the beginning and work outward towards your goal. You may come to a dead end, which means back-tracking or possibly starting over. We are going to call this **forward chaining** (chain, as we are making a path of links from leaf to leaf in the tree).

- The other way is to start with the goal and work up the tree toward the start. This is **backward chaining**. The cool thing about backward chaining is that there are a lot fewer branches to traverse. You can guess that a major problem with backward chaining is you have to know what all the leaves are in advance before you can use them. In many problems, such as a grid search or a path planner, this is possible. It does not work in chess, with an exponentially massive tree.

- The third technique? No one says we can't do both – we could combine both forward and backward chaining and meet somewhere in the middle.

The choice of decision tree shapes, chaining techniques, and construction is based on the following:

- What data is available?

- What information is known or unknown? How is the path scored or graded?

There are also different kinds of solutions for path planning using decision trees. If you were given unlimited resources, the biggest computer, perfect knowledge in advance, and are willing to wait, then you can generate an **optimal path** or solution.

One of my lessons learned from years of developing practical AI-based robots and unmanned vehicles is that any solution that meets all of the criteria or goals is an acceptable and usable solution, and you don't have to wait and continue to compute the perfect or optimal solution. Often then, a *good enough* solution is found in 1/10 or even 1/100 the time of an optimal solution, because an optimal solution requires an exhaustive search that may have to consider all possible paths and combinations.

So, how do we approach making our decision trees work faster, or more efficiently? We do what any good gardener would do – start pruning our trees.

What do we mean by pruning?

Sometimes in the computer business, we have to make metaphors to help explain to people how something works. You may remember the desktop metaphor that Apple, and later, Windows, adopted to help explain graphical operating systems. Sometimes, we just run those metaphors into the ground, such as the trash can to delete files, or *Clippy*, the paper clip assistant.

You may feel that I've gone off the metaphorical deep end when I discuss **pruning** your decision trees. What's next, fertilizer and tree spikes? Actually, pruning is a critical concept in decision tree-type systems. Each branch in your tree can lead to hundreds or thousands of sub-branches. If you can decide early that a branch is not useful, you can cut it out and you don't have to process any of the branches or leaves in that branch. The sooner you can discover that a path is not getting you to your goal, the quicker you can reduce the time and effort involved in creating a solution, which is a real-time system such as a robot, a self-driving car, or an autonomous aircraft; this can spell the difference between usable and worthless.

Let's run through a quick example in which we use the pruning method. One great use for a decision tree process is **Fault Detection, Isolation, and Recovery (FDIR)**. This is a typical function of a robot. Let's make a decision tree for FDIR in the case of our Tinman robot not moving. What automated steps could we take to detect the fault, isolate the problem, and then recover? One technique we can use is **root cause analysis,** where we try to figure our problem by systematically listing and then eliminating (pruning) causing factors and seeing whether the symptoms match. One way to approach root cause analysis is to use a special form of decision tree called a **fishbone diagram**, or **Ishikawa diagram**. This diagram is named after its inventor, Professor Kaoru Ishikawa from the University of Tokyo. In his 1968 paper, *Guide to Quality Control,* the fishbone diagram is named because of its shape, which has a central spine and ribs jutting off on either side. I know, the metaphors are getting deep when we have a decision *tree* in the shape of a *fish.*

Now, we begin to have a problem. Remember that in a robot, a problem is a symptom, not a cause. Our problem is the robot is not moving. What can cause this problem? Let's make a list:

- The drive system

- The software

- The communication system

- The battery and wiring

- The sensors

- Operator error

Now, for each of these, we subdivide our branches into smaller branches. What parts of the *drive system* can cause the robot to not be able to move? The wheels could be stuck. The motors could not be getting power. The gears could be jammed. The motor driver could have overheated. Here is my fishbone diagram to illustrate the problem of the robot not moving:

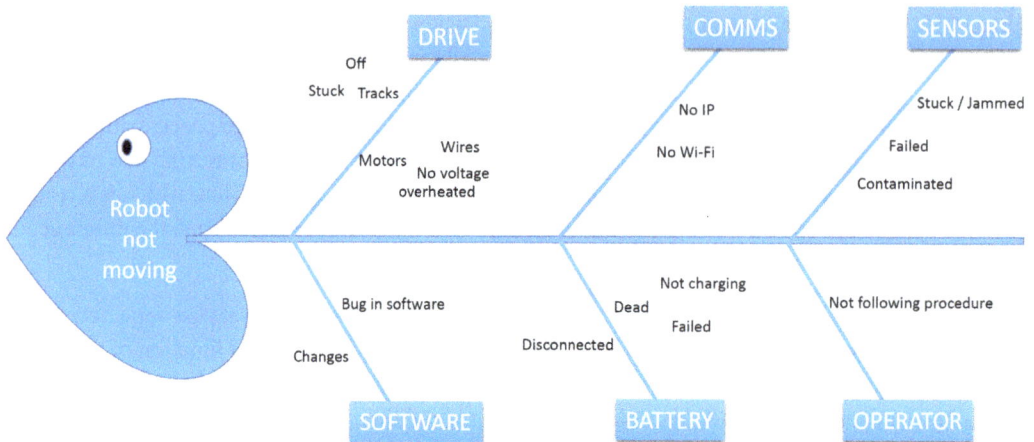

Figure 8.2 – A fishbone, or Ishikawa, diagram is commonly used for troubleshooting

For each of these factors, you can consider what would be the symptoms of that problem being the cause. If the gears in the motors are jammed, then the motors can't turn and the wheels can't turn. If we can check any of these factors off, we can prune or eliminate the gears from our diagram or decision tree. We check the gears, and the wheels and motors turn by hand, so the gears are not the cause. We prune that branch. If we have an automated way of doing testing, we can automatically prune branches, which we will be able to do in the later examples in this chapter.

How about the battery? The battery could need charging (dead battery), the battery could be disconnected, or a power wire could be loose. We check the battery voltage – that is OK, so prune that leaf off the tree. We check the wiring – nothing loose. The battery branch gets pruned.

And so we go on until we have something that either matches all our symptoms or is the last one left. Let's say the last branch was communications. Now what? We ask, "What things in communications would cause us not to move?" Our first answer is that motor command messages are not getting through to our robot over the network. We check the log and see, indeed, no motor messages are present (cmd_vel, in our case). There is our problem, but what caused the problem? The network could be broken (checked – no, the network is OK), or the IP address could be wrong (no, that's OK). We look to see whether any recent changes were made to the control software, and indeed, there were. We revert to the previous version and see the robot move. There is our problem and we used a decision tree to find it.

So, in this case, we solved our problem almost entirely by pruning branches and leaves off our tree until only one path was left, or we arrived at our goal.

How can we prune branches in software? We can look for *dead ends*. Dead ends are leaves – parts of the tree that end and have no future branches. When we reach a dead end, we can not only prune that leaf but also the parts of the path that exclusively lead to that branch. This would be a **backward-chaining** approach to pruning, as we start at the end and work backward.

We can also see sections of the tree that are unused, or never referenced or called. We can remove entire sections in this manner. This is **forward chaining** because we are traversing the tree in the forward direction, from the front to the back.

Up to this point, we, the humans in the story, have been making these decision trees by hand. We have not even discussed how we write a program to allow the robot to use trees to make decisions. Wouldn't it be a lot nicer if the computer was doing all the hard work of making the tree, deciding the branches, and labeling the nodes instead of us? That is exactly what we will discuss in the next section.

Creating self-classifying decision trees

Let's consider the problem of classifying toys. We may want to come up with a more efficient robot, which sorts toys in some manner instead of just dumping them in a box. In an ideal world, out of a population of 20 toys, we would have some characteristics that divided the group evenly in half – 10 and 10. Let's say it is length – half of the toys are under six inches long and half are over. Then, it would also be ideal if some other characteristic divided each of those groups of 10 in half – into four groups of five.

Let's say it's *color* – we have five red toys, five blue toys, five green toys, and five yellow toys. You may recognize that we are doing what biologists do in classifying new species – we are creating a **taxonomy**. Now, we pick another attribute that separates the toys into even smaller groups – it might be what kind of toy it is or what size wheels it has. I think you get the picture. Let's look at an example.

Now, what would be great is if we could list all the toys and all the attributes in a table, and let the computer figure out how many groups and what kinds there are. We could create a table like this one:

Type	Length	Width	Weight	Color	Number of wheels	Noise	Soft	Material	Eyes	Toy Name
car	3	1	35	red	4	0	hard	metal	0	hotwheels
car	3	1	35	orange	4	0	hard	metal	0	hotwheels
car	3	1	35	blue	4	0	hard	metal	0	hotwheels
car	3	1	35	blue	4	0	hard	metal	0	hotwheels
car	3	1	35	white	4	0	hard	metal	0	hotwheels
stuffed	5	5	50	white	0	0	verysoft	fur	2	plush

Type	Length	Width	Weight	Color	Number of wheels	Noise	Soft	Material	Eyes	Toy Name
stuffed	7	5	55	brown	0	0	verysoft	fur	3	plush
action	2	4	80	gray	0	0	hard	metal	0	slinky
build	2	2	125	wood	0	0	hard	wood	0	wood block 2x2
build	2	2	75	wood	0	0	hard	wood	0	wood block triangle
build	4	2	250	wood	0	0	hard	wood	0	wood block 4x2
dish	3	3	79	blue	0	0	hard	ceramic	0	teapot
aircraft	7	5	65	white	4	1	hard	plastic	0	space shuttle
aircraft	13	7	500	green	8	1	hard	plastic	0	Thunderbird 2
car	5	1	333	yellow	6	1	hard	metal	0	school bus
music	12	4	130	wood	0	2	hard	wood	0	toy guitar
music	5	2	100	yellow	0	1	hard	plastic	0	play microphone
music	4	4	189	white	0	2	hard	wood	0	toy drum

Table 8.1 – A table of attributes for a group of toys used for classification

We now have a problem we have to solve. We will be using a decision tree classifier that is provided with the `scikit-learn` Python package called `DecisionTreeClassifier`. This program cannot use strings as input data. We will have to convert all of our string data into some sort of numeric figure. Fortunately, the `scikit-learn` library provides us with a function just for this purpose. It provides several encoding functions that convert strings into numbers. The function we will use is called `LabelEncoder`. This function takes an array of strings and converts it into an enumerated set of integers.

We can take our first column, which has the type of toy. My nomenclature is *toy = toy car, stuffed = stuffed animal, aircraft = toy aircraft*, and *music = toy musical instrument*. We also have *action* for *action toy*, and *build* for *building toy* (that is, blocks, LEGO™, and so on). We'll have to turn these into some sort of numbers.

LabelEncoder will convert a column in our data table that is populated with strings. The `type` column from the data is shown in the following code:

```
['car' 'car' 'car' 'car' 'car' 'stuffed' 'stuffed' 'action' 'build'
'build' 'build' 'dish' 'aircraft' 'aircraft' 'car' 'music' 'music'
'music']
```

It converts it to the label-encoded toy type:

```
[3 3 3 3 3 6 6 0 2 2 2 4 1 1 3 5 5 5]
```

You can see that everywhere where it said `car`, we now have the number 3. You can also see that 6 = `stuffed`, 0 = `action`, and so on. Why the odd numbering? The encoder first sorts the strings in alphabetical order.

We are going to just dive right in from here to create a classification program:

1. Here is our decision tree classifier program:

    ```
    # decision tree classifier
    # author: Francis X Govers III #
    # example from book "Artificial Intelligence for
    Robotics" #
    ```

2. We first import the libraries we will be using. There is an extra library called `graphviz` that is useful for drawing pictures of decision trees. You can install it with the following:

    pip install graphviz

3. We are going to be using the `pandas` package, which provides a lot of data table-handling tools:

    ```
    from sklearn import tree
    import numpy as np
    import pandas as pd
    import sklearn.preprocessing as preproc
    import graphviz
    ```

4. Our first step is to read in our data. I created my table in Microsoft Excel and exported it as a **comma-separated values** (CSV) format. This allows us to read in the data file directly with the column headers. I print out the shape and size of the data file for reference. My version of the file has 18 rows and 11 columns. The last column is just a note to myself on the actual name of each toy. We will not be using the last column for anything. We are building a classifier that will separate the toys by type:

```
toyData = pd.read_csv("toy_classifier_tree.csv")
print ("Data length ",len(toyData))
print ("Data Shape ",toyData.shape)
```

5. Now, we can start building our decision tree classifier. We first build an instantiation of the `DecisionTreeClassifer` object. There are two different types of **decision tree classification** (DTC) algorithms to choose from:

    • **Gini coefficient**: The Gini coefficient was developed in 1912 by the Italian statistician Corrado Gini in his paper, *Variabilita e Mutabilita*. This coefficient, or index, measures the amount of inequality in a group of numbers. A zero value means all the members of the group are the same.

    • **Entropy method**: Entropy, when we are talking about AI, refers to the amount of uncertainty in a set of data. This concept comes from information theory, in which it measures the amount of uncertainty in a random variable. The concept was introduced by Claude Shannon in the 1940s. To create a decision tree, the algorithm tries to decrease entropy (reduce uncertainty) by splitting the group at a point where each child node is more homogenous than its parent.

    Here, we are going to use the Gini coefficient. If we had a group of toy cars that were all the same size and all red, then the Gini coefficient of the group would be 0. If the members of the group are all different, then the Gini coefficient is closer to 1. The Gini coefficient is given by the following equation:

    $$G(S) = 1 - \sum i=1 \, n \, p \, i \, 2$$

    We have 4 toy cars out of 18 toys, so the probability of a toy car being in a group is *4/18* or 0.222. The decision tree will continue to subdivide classes until the Gini coefficient of the group is 0:

```
dTree = tree.DecisionTreeClassifier(criterion ="gini")
```

6. We need to separate out the values in our data table. The data in the first column, which is called column 0 in Python, are our classification labels. We need to pull those out separately, as they are used to separate the toys into classes. From our previous work with neural networks, these would be our outputs or the label data we have used in other machine learning processes. We will be training our classifier to predict the class of the toy based on the attributes in the table (size, weight, color, and so on). We use slicing to pull the data out of the pandas table. Our

pandas data table is called `toyData`. If we want the entries in the table, we need to ask for `toyData.values`, which will be returned as a 2D array:

```
dataValues=toyData.values[:,1:10]
classValues = toyData.values[:,0]
```

If you are not familiar with slicing notation in Python, the statement `toyData.values[:,1:10]` returns just the columns in our table from 1 to 10 – it leaves column 0 out. We actually have 11 columns in our table, but since Python starts numbering them at 0, we end up needing 1 to 10. You will probably guess that the other notation just grabs the data in the first column.

7. This is the label encoder that we talked about – it will convert the strings in our data into numbers. For example, colors such as *red*, *green*, and *blue* will be converted to numbers such as *0*, *1*, and *2*. The first item to be encoded is the list of class values that we use to label the data. We use the `LabelEncoder.fit()` function to come up with the formula for converting strings to numbers, and then the `LabelEncoder.transform()` function to apply it. Note that `fit()` does not produce an output.

8. Finally, we need to make the string text and the list of encoded numbers match up. What `LabelEncoder` will do is sort the strings alphabetically and start numbering them from *A*, ignoring any duplicates. If we put in `car, car, car, block, stuffed, airplane`, we will get `2,2,2,1,3,0` as the encoding, and we will have to know that `airplane = 0`, `block = 1`, `car = 2`, and `stuffed = 3`. We need to generate a **decoder ring** to match up the numbers and text descriptions that look like `airplane, block, car, stuffed`. We duplicate the `LabelEncoder` function by using two functions on our list of string-formatted class names:

    -   We use the `set()` function to eliminate duplicates

    -   We use the `sorted()` function to sort in the correct order

    Now, our class name table and the enumerations generated by `LabelEncoder` match. We'll need this later:

```
lencoder = preproc.LabelEncoder() lencoder.
fit(classValues)
classes = lencoder.transform(classValues)
classValues = list(sorted(set(classValues)))
```

9. To make it easy on ourselves, I created a function to automatically find out which columns in our data are composed of strings and to convert those columns into numbers. We start by building an empty list to hold our data. We will iterate through the columns in our data and look to see whether the first data value is a string. If it is, we will convert that whole column into numbers using the label encoder object (`lencoder`) we created. The label encoding

process has two parts. We call `lencoder.fit()` to see how many unique strings we have in our column and to create a number for each one. Then, we use `lencoder.transpose` to insert those numbers into a list:

```
newData = []
for ii in range(len(dataValues[0]))
line = dataValues[:,ii]
if type(line[0])==str:
    lencoder.fit(line)
  line = lencoder.transform(line)
```

10. Now, we put all of the data back into the `newData` list, but there is a problem – we have turned all our columns into rows! We use the `transpose` function from `numpy` to correct this problem. But wait! We don't have an array anymore, as we turned it into a list so we could take it apart and put it back together again (you can't do that with a `numpy` array – believe me, I tried):

```
newData.append(line)
newDataArray = np.asarray(newData)
newDataArray = np.transpose(newDataArray)
```

11. Now, all of our preprocessing is done, so we can finally call the real `DecisionTreeClassifer`. It takes two arguments:

- The array of our data values

- The array of class types that we want the decision tree to divide our groups into

`DecisionTreeClassifier` will determine what specific data from the table is useful for predicting what class one of our toys fits into:

```
dTree = dTree.fit(newDataArray,classes)
```

That's it – one line. But wait – we want to see the results. If we just try and print out the decision tree, we get the following:

```
DecisionTreeClassifier(class_weight=None,
criterion='gini', max_depth=None, max_features=None, max_
leaf_nodes=None,
min_impurity_split=1e-07, min_samples_leaf=1, min_
samples_split=2, min_weight_fraction_leaf=0.0,
presort=False, random_state=None, splitter='best')
```

That does not tell us anything; that is a description of the `DecisionTreeClassifier` object (it does show us all of the parameters we can set, which is why I put it here).

12. So, we use a package called `graphviz`, which is very good at printing decision trees. We can even pass our column names and class names into the graph. The final two lines output the graph as a `.pdf` file and store it on the hard drive:

```
c_data=tree.export_graphviz(dTree,out_file=None,feature_
names=toyData.colum ns, class_names=classValues, filled =
True, rounded=True,special_characters=True)
graph = graphviz.Source(c_data)
graph.render("toy_graph_gini")
```

And here is the result. I will warn you, this is addictive:

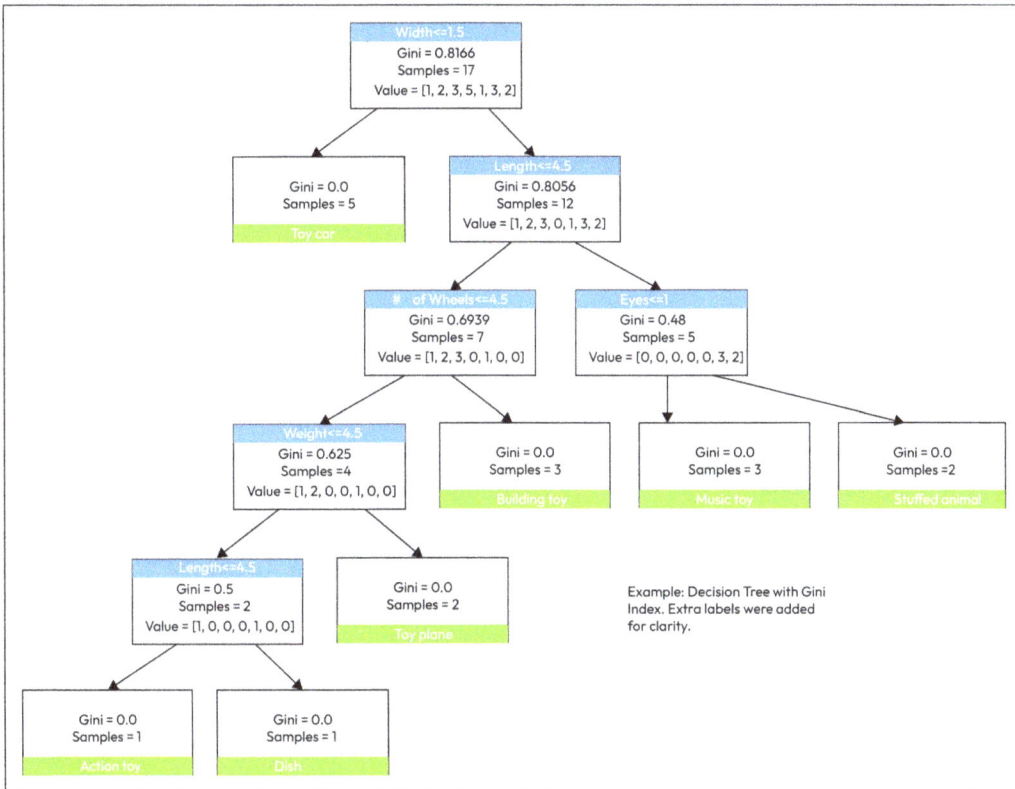

Figure 8.3 – The output of the decision tree using the Gini index method

We can quickly check our solution by looking at our input table and seeing whether the numbers line up. We should see the following:

- Five toy cars
- Three building blocks
- One dish
- One action toy
- Two stuffed animals
- Three musical instruments
- Two toy airplanes

And that is indeed the case.

The other number to look at is the Gini index. As shown in *Figure 8.3*, the top-level box shows that the index for the entire group has an overall value of 0.8166, which is close to 1 and shows a high degree of heterogeneity. As we progress down the tree, the Gini numbers get smaller and smaller until reaching 0 at each of the identified groups, which shows that the items in those groups share all of the same attributes.

What does this graph tell us? First of all, we can separate the toy cars by only one attribute – *width*. Only the toy cars are less than 1.5 inches wide (38 mm). We don't need to look at color, weight, or anything other than width to separate all the toy cars from everything else. We see we have 5 toy cars out of our 18 toys, so we have 13 left to classify. Our next division comes in length. We have 7 toys less than 4.5 inches long (11 cm) and 5 that are longer. Of the group of five, two have eyes and three do not. The toys with eyes are the two stuffed animals. If you follow the tree, the branches that lead to the toy music instruments are width > 1.5 inches, length > 4.5 inches, and no eyes, and they are indeed larger than the other toys in length and width, and don't have eyes.

None of the other bits matter in terms of classifying. That means that an attribute such as *color* is a poor predictor of what class a toy belongs to – which makes sense. Our other useful criteria are the *number of wheels*, the *weight*, and the *length*. That data is sufficient to classify all our toys into groups. You can see that the Gini index of each leaf node is indeed 0. I added some additional labeling to the graph to make the illustration clearer, as the program uses the class number rather than the class name in the graph.

So, that exercise was satisfactory – we were able to create an automatic decision tree from our toy data that classified our toys. We can even use that data to classify a new toy and predict which class it might belong to. If we found that that new toy violated the classification somehow, then we would need to re-rerun the classification process and make a new decision table.

There is another type of process for creating decision trees and subdividing data into categories. That is called the **entropy model**, or **information gain**. Let's discuss this next.

Understanding entropy

Entropy is a measurement of the amount of disorder in the sample of data provided. We can also call this process **information gain** since we are measuring how much each criterion contributed to our knowledge of which class it belongs to.

The formula for entropy is a negative log base 2 function that is still primarily looking at the probability of a class belonging to a population, which is just the number of individuals belonging to each class divided by the total number in the sample:

$$Entropy = -p*log2(p) - p\_i*log2(p\_i)$$

To substitute entropy as our group criteria in our program, we only have to change one line:

```
dTree = tree.DecisionTreeClassifier(criterion ="entropy")
```

The results are shown in the following diagram:

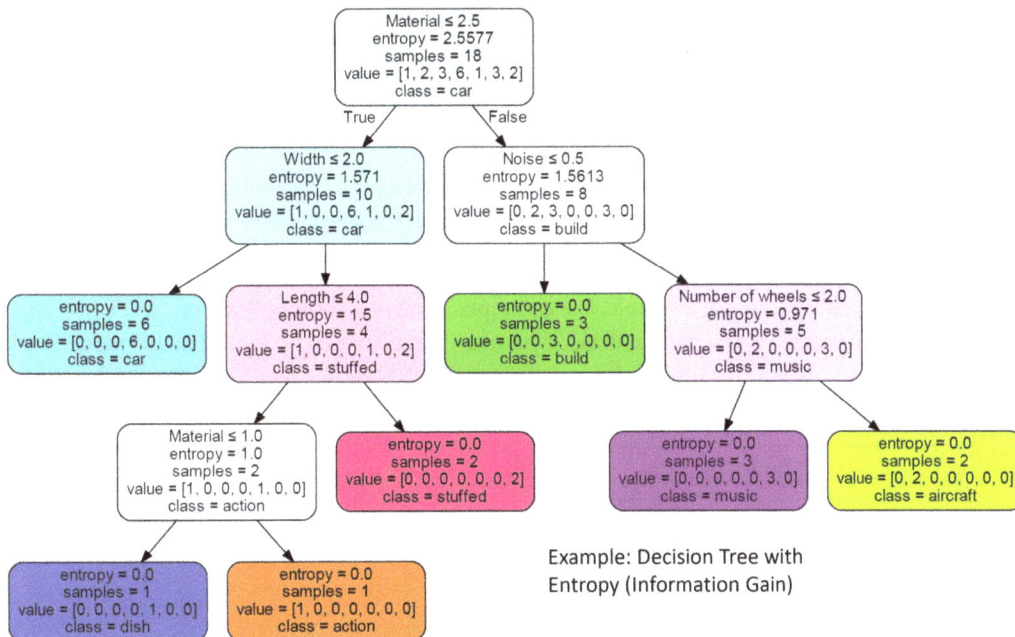

Example: Decision Tree with Entropy (Information Gain)

Figure 8.4 – Output of the decision tree using entropy (information gain)

You can note that entropy starts at 2.55 for our whole group, and decreases to 0 at the leaf nodes (ends of the branches). We can check that we have seven classifications, but you can see that the entropy method selected different criteria from the Gini method. For example, the Gini classifier started with `Length`, and the entropy classifier started with `Material`. The entropy method also chose

Noise (whether the toy makes a noise or not) and correctly selected that the only toys that make a noise were the toy musical instruments and the toy airplanes, which have electronic sound boxes that make airplane sounds.

There is one item that causes some concern, however. There are two blocks that show Material, dividing the toy's values in material less than 2.5. Material is a discrete value. We can generate a list of materials and run this through our sorted(set(list)) process to get the unique values in sorted order:

```
['ceramic', 'fur', 'metal', 'plastic', 'wood']
```

So, a Material value of 2.5 or less would be either ceramic or fur. Fur and ceramic have nothing in common, other than where they are found in the alphabet. This is a rather troubling relationship, which is an artifact of how we encoded our data as a sequential set of numbers. This implies relationships and grouping that don't really exist. How can we correct this?

As a matter of fact, there is a process for handling just this sort of problem. This technique is widely used in AI programs and is a *must-have* tool for working with classification, either here in the decision tree section or with neural networks. This tool has the strange name of **one-hot encoding**.

Implementing one-hot encoding

The concept for one-hot encoding is pretty simple. Instead of replacing a category with an enumeration, we add one column to our data for each possible value and set it to be a 1 or 0 based on that value. The name comes from the fact that only one column in the set is *hot* or selected.

We can apply this principle to our example. We can replace the one column, Material, with five columns for each material type in our database: ceramic, fur, metal, plastic, and wood:

Material	ceramic	fur	metal	plastic	wood
metal	0	0	1	0	0
metal	0	0	1	0	0
metal	0	0	1	0	0
metal	0	0	1	0	0
metal	0	0	1	0	0
fur	0	1	0	0	0
fur	0	1	0	0	0
metal	0	0	1	0	0

Material	ceramic	fur	metal	plastic	wood
wood	0	0	0	0	1
wood	0	0	0	0	1
wood	0	0	0	0	1
ceramic	1	0	0	0	0
plastic	0	0	0	1	0
plastic	0	0	0	1	0
metal	0	0	1	0	0
wood	0	0	0	0	1
plastic	0	0	0	1	0
wood	0	0	0	0	1

Table 8.2 – One-hot encoding data structure for the Material category

This does cause some structural complications to our program. We must insert columns for each of our types, which replaces 3 columns with 14 new columns.

I've found two functions that we can use to convert text categories into one-hot encoded multiple columns:

- One is OneHotEncoder, which is part of scikit-learn. It is used like LabelEncoder – in fact, you must use both functions at the same time. You have to convert the string data to numeric form with LabelEncoder and then apply OneHotEncoder to convert that to the one-bit-per-value form that we want.

- The simpler way is with a pandas function called get_dummies(). The name is apparently because we are creating dummy values to replace a string with numbers. It does perform the same function. The steps involved are quite a bit simpler than using the OneHotEncoder process, so that will be the one in our example.

Let's look at the steps we need to follow to implement this:

1. The top header section is the same as before – we have the same imports:

```
# decision tree classifier
# with One Hot Encoding and Gini criteria #
# Author: Francis X Govers III #
# Example from book "Artificial Intelligence for
Robotics" #
```

```
from sklearn import tree
import numpy as np
import pandas as pd
import sklearn.preprocessing as preproc
import graphviz
```

2. We will begin by reading in our table as before. I added an extra column at my end called Toy Name so I could keep track of which toy is which. We don't need this column for the decision tree, so we can take it out with the pandas del function by specifying the name of the column to remove:

```
toyData = pd.read_csv("toy_classifier_tree.csv")
del toyData["Toy Name"]     # we don't need this for now
```

3. Now, we are going to create a list of the columns we are going to remove and replace from the pandas dataTable. These are the Color, Soft, and Material columns. I used the term *Soft* to identify toys that were soft and squished easily (as compared to hard plastic or metal) because that is a separate criterion we may need for using our robot hand. We generate the dummy values and replace the 3 columns with 18 new columns. pandas automatically names the columns with a combination of the old column name and the value. For example, the single Color column is replaced by Color_white, Color_blue, Color_green, and so on:

```
textCols = ['Color','Soft','Material']
toyData = pd.get_dummies(toyData,columns=textCols)
```

4. I put a print statement here just to check that everything got assembled correctly. It is optional. I've been really impressed with pandas for data tables – there is a lot of capability there to do database-type functions and data analysis:

```
print toyData
```

5. Now, we are ready to generate our decision tree. We instantiate the object and call it dTree, setting the classification criteria to Gini. We then extract the data values from our toyData dataframe, and put the class values in the first (0th) column into the classValues variable, using array slicing operators:

```
dTree = tree.DecisionTreeClassifier(criterion ="gini")
dataValues=toyData.values[:,1:]
classValues = toyData.values[:,0]
```

6. We still need to convert the class names into an enumerated type using `LabelEncoder`, just as we did in the previous two examples. We don't need to one-hot encode. Each class represents an end state for our classification example – the leaves on our decision tree. If we were doing a neural network classifier, these would be our output neurons. One big difference is that when using a decision tree, the computer tells you what the criteria were that it used to classify and segregate items. With a neural network, it will do the classification but you have no way of knowing what criteria were used:

```
lencoder = preproc.LabelEncoder()
lencoder.fit(classValues)
classes = lencoder.transform(classValues)
```

7. As we said, to use the class value names in the final output, we have to eliminate any duplicate names and sort them alphabetically. This pair of nested functions does that:

```
classValues = list(sorted(set(classValues)))
```

8. This is the conclusion of our program. Actually creating the decision tree only takes one line of code, now that we have set up all the data. We use the same steps as before, and then create the graphic with `graphviz` and save the image as a PDF. That was not hard at all – now that we have had all that practice setting this up:

```
print ""
dTree = dTree.fit(dataValues,classes)
c_data=tree.export_graphviz(dTree,out_file=None,feature_
names=toyData.columns,
class_names=classValues, filled = True,
rounded=True,special_characters=True)
graph = graphviz.Source(c_data) graph.render("toy_
decision_tree_graph_oneHot_gini")
```

The result is the flowchart shown in the following figure. This output with one-hot encoding is a bit easier to read than *Figure 8.4* because we can see the numbers in each category. You'll note that each leaf (end node) has only one category with a count (two stuffed animals and three musical instruments):

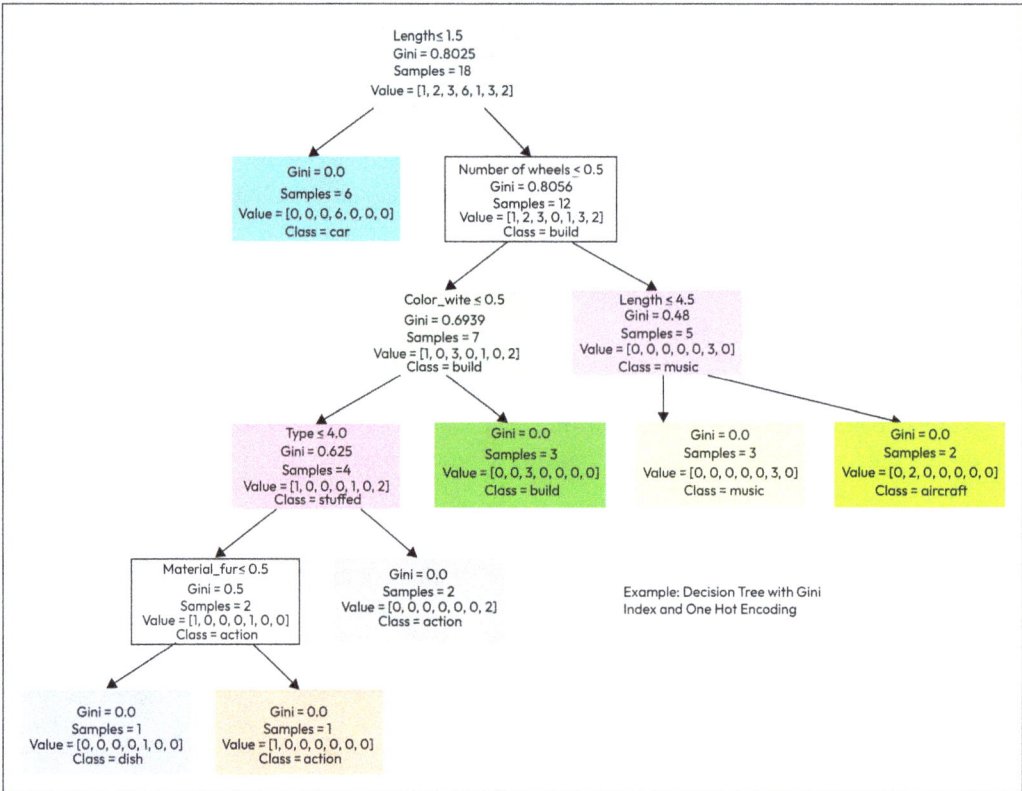

Figure 8.5 – The output of the decision tree using one-hot encoding is much easier to read

Since we've been able to describe and make all sorts of decision trees, what would we have if we used a whole bunch of them? A forest! Let's explore what this might look like.

Random forests

I really wanted to add this section on **random forest classifiers**, but not just because the name sounds so cool. While I may have been accused of stretching metaphors to the breaking point, this time, the name may have inspired the name of this type of decision tree process. We have learned how to make decision trees, and we have learned that they have some weak points. It is best if the data really belongs to distinct and differentiated groups. They are not very tolerant of noise in the data. And they really gets unwieldy if you want to scale them up – you can imagine how big a graph would get with 200 classes rather than the 6 or 7 we were dealing with.

If you want to take advantage of the simplicity and utility of decision trees but want to handle more data, more uncertainty, and more classes, you can use a random forest, which, just as the name indicates, is just a whole batch of randomly generated decision trees. Let's step through the process:

1. We collect our database of information but, instead of 18 rows in our database, we have 10,000 records or 1 million records. We subdivide this data into random sets – we generate 100 sets of data each *randomly* chosen from all of our data – and we put them in *random* order. We also pull out one set of data to use as a test set, just as we did for the neural networks.

2. Now, for each set of random data, we make a decision tree using the same process we have already learned.

3. Now, we have this collection of 100 classification engines, each generated from a different, randomly generated subset of data. We now test our random forest by taking data from the test set and running through all 100 of the trees in our forest. Each tree will provide an estimate of the classification of the data in our test record. If we are still classifying toys, then one of the trees would estimate that we are describing a toy car. Another may think it's a musical instrument. We take each estimate and treat it as a vote. Then, the majority rules – the class that the majority of the trees selected is the winner. And that is all there is to it.

The setup and program are just the same as what we did before, but you can't draw a decision tree from a random forest, or just create a tree as an end in itself because that is not what a random forest does – if you just need a decision tree, you know how to do that. What you can do is to use a random forest like a neural network, as either a classification engine (to what class does this data belong?) or a regression engine that approximates a non-linear curve.

At this point, you can conclude with me that decision trees are really useful for a lot of things. But did you know you can navigate with them? The next section covers path planning for robots – using a different type of decision tree.

Introducing robot path planning

In this section, we will be applying decision tree techniques to perform robot navigation. Some people like to refer to these as **graph-based solutions**, but any sort of navigation problem ends up being a decision tree. Consider as you drive your car, can you divide your navigation problems into a set of decisions – turn right, turn left, or go straight?

We are going to take what we have learned so far and press on to a problem related to classification, and that is **grid searching** and **path finding**. We will be learning about the famous and widely used **A*** (pronounced **A-star**) algorithm. This will start with grid navigation methods, topological path finding, such as GPS route finding, and finally, expert systems. You will see that these are all versions and variations on the topic of decision trees that we have already learned.

Some problems and datasets, particularly in robotics, lend themselves to a grid-based solution as a simplification of the navigation problem. It makes a lot of sense that, if we were trying to plot a path around a house or through a field for a robot, we would divide the ground into some sort of checkerboard grid and use that to plot coordinates that the robot can drive to. We could use latitude and longitude, or we could pick some reference point as zero – such as our starting position – and measure off some rectangular grid relative to the robot. The grid serves the same purpose in chess, limiting the number of positions under consideration for potential future movement and limiting and delineating our possible paths through the space.

While this section deals with gridded path finding, regardless of whether maps are involved or not, there are robot navigation paradigms that don't use maps and even some that don't use grids, or use grids with uneven spacing. I've designed robot navigation systems with multiple-layer maps where some layers were mutable – changeable – and some were not. This is a rich and fertile ground for imagination and experimentation, and I recommend further research if you find this topic interesting. For now, let's start with a description of the coordinate system we'll be using.

Understanding the coordinate system

Let's get back to the topic at hand. We have a robot and room that is roughly rectangular, and within that rectangle are also some roughly rectangular obstacles in the form of furniture, chairs, bookcases, a fireplace, and so on. It is a simple concept to consider that we mark off a grid to represent this space and create an array of numbers that matches the physical room with a virtual room. We set our grid spacing at 1 cm – each grid square is 1 cm x 1 cm, giving us a grid with 580 x 490 squares or 284,200 squares. We represent each square by an unsigned integer in a 2D array in the robot's memory.

Now, we are going to need some other data. We have a starting location and a goal location, specified as grid coordinates. We'll put 0, 0 for the grid in the nearest and leftmost corner of the room so that all our directions and angles will be positive. In the way I've drawn the room map for you in *Figure 8.6*, that corner will always be the lower-left corner of our map. In standard *right-hand rule* notation, left turns are positive angles and right turns are negative. The x direction is horizontal and the y direction is vertical on the page. For the robot, the x axis is out the right side and the y axis is the direction of motion.

You may think it odd that I'm giving these details, but setting up the proper coordinate system is the first step in doing grid searches and path planning. We are using Cartesian coordinates indoors. We would use different rules outdoors with latitude and longitude. There, we might want to use *north-east-down* (north is positive, south is negative, east is positive, west is negative, the z axis is down, and the x axis is aligned on the robot with the direction of travel):

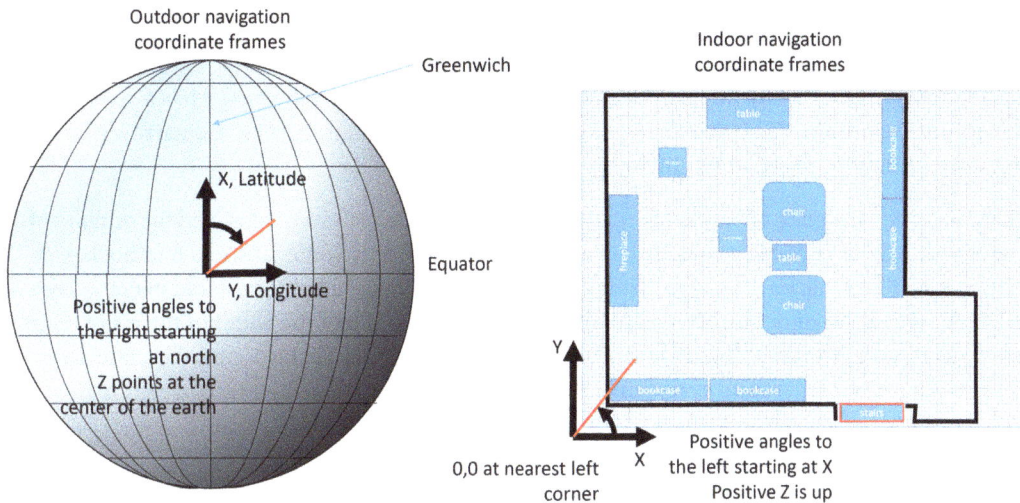

Figure 8.6 – Coordinate frames for Earth navigation and indoor navigation

We will be looking at this room map in more detail later.

So, we have our grid and a coordinate system that we agree upon, or at least agree that we both understand. We also have a starting location and an ending location. Our objective is to determine the best path for the robot from the start to the finish point. And in between, we have to plan a path around any obstacles that may be in the way.

Next, we have to talk about knowledge.

Developing a map based on our knowledge

There are basically two kinds of grid search and path finding routines:

- **A priori knowledge**, where you know where everything is on the map
- **A posteriori knowledge**, where you don't know where the obstacles are

We will start in the easier position where we can do our path planning with perfect knowledge of the layout of the room – we have a map.

We really have three goals we are trying to achieve simultaneously with path planning:

- Reach our goal
- Avoid obstacles
- Take the shortest path

We can talk about how we might go about this. We can start with our pencil at the start point and draw an imaginary line from our start to the goal. If there are no obstacles in the way, we are done. But wait – our pencil is a tiny line on paper. Our robot is somewhat chubbier – it has a significant width as it drives around. How do we judge whether the robot is going down some narrow passage that it won't fit into? We need to modify our map!

We have our grid, or a piece of paper that represents the grid. We can draw on that grid the outlines of all the obstacles, to scale. We have two chairs, two tables, a fireplace, two ottomans, and four bookcases. We color in all the obstacles in the darkest black we can. Now, we get a lighter colored pencil – say a blue color – and draw an outline around all of the furniture that is half the width of the robot. Our robot is 32 cm wide, so half of that is 16 cm, a nice even number. Our grid is 1 cm per square, so we make a 16-square border around everything. It looks like this:

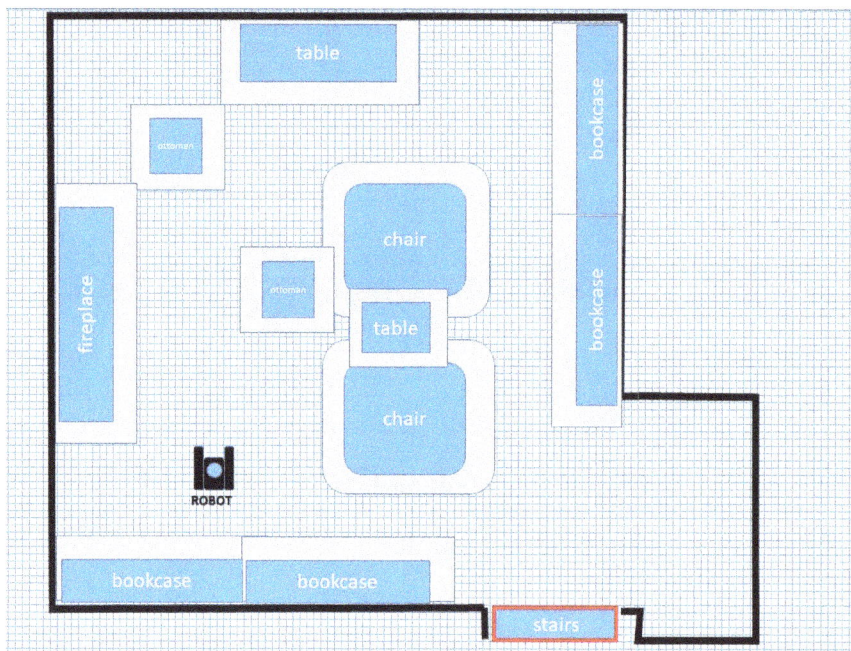

Figure 8.7 – Adding safety boundaries to obstacles helps prevent collisions

So, now our map has two colors – obstacles and a *keep-out* border. We are going to keep the center of the robot out of the keep-out zone, and then we will not hit anything. This should make sense. As for judging passages and doorways, if the keep-out zones touch on either side (so if there are no white squares left in the middle), then the robot is too big to pass. You can see this around the ottoman in the upper-left corner of the illustration.

We look at our line now. We need a way to write a computer algorithm that determines the white squares that the robot can pass through that gets us from the start point to the finish point.

Since we have the goal in Cartesian coordinates and we have our start spot, we can express the distance in a straight line from the start to the finish. If the start point is x1, y1 and the finish point is x2, y2, then the distance is the square root of the sums of the difference between the points:

$$distance = sqrt(x2-x1)^2 + (y2-y1)^2)$$

One approach for developing a path planning algorithm is to use a **wavefront method**. We know where the start is. We go out in every direction to the eight squares adjacent to the start point. If any of those hit an obstacle or keep-out zone, we throw it out as a possible path. We keep track of how we got to each square, which, in my illustration (*Figure 8.8*), is indicated by the arrows. We use the information on how we got to the square because we don't yet know where we are going next. Now, we take all the new squares and do the same thing again – grabbing one square, seeing which of its eight neighbors is a legal move, and then putting an arrow (or a pointer to the location of the previous square) in it to keep track of how we got there. We continue to do this until we get to our goal. We keep a record of the order of the squares we examined and follow the arrows backward to our starting point.

If more than one square has a path leading to the current square, then we take the closest one, which is to say the shortest path. We follow these predecessors all the way back to the starting point, and that is our path:

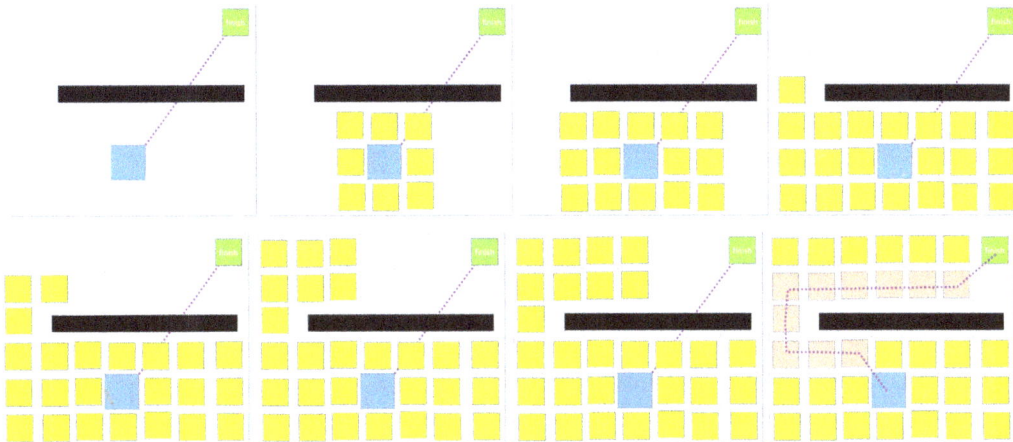

Figure 8.8 – The wavefront approach to path planning has very little math involved. Each figure is a step in the process, starting at the upper left and going across, then down

You will notice in this example that I allowed the robot to make diagonal turns to get from one square to another. I could have also specified that only right-angle turns are allowed, but that is not very efficient and is hard on the robot's drive system. Only allowing right-angle turns simplifies the processing somewhat, since you only have to consider four neighbors around a square instead of eight.

Another approach for developing a path planning algorithm that would look promising is the **Greedy Best-First** approach. Instead of keeping a record and checking all of the grid points as we did in the wavefront method, we just keep the single best path square out of the eight we just tested. The measure we use to decide which square to keep is the one that is closest to our straight-line path. Another way of saying this is to say it's the square that is closest to the goal. We remove squares that are blocked by obstacles, of course. The net result is we are considering a lot (really a lot!) fewer squares than the wavefront method of path planning:

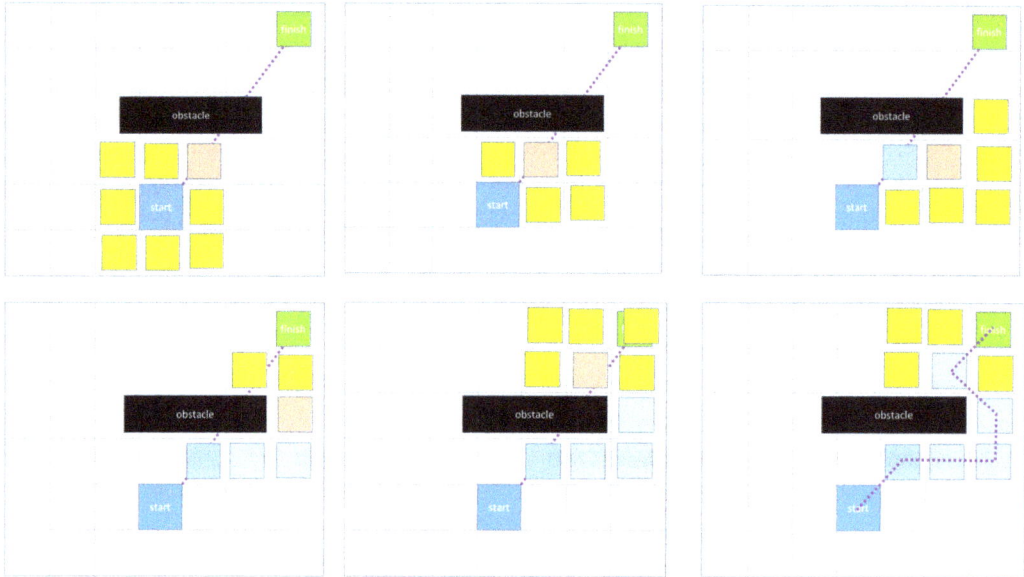

Figure 8.9 – The aptly named "Greedy Best-First" algorithm is fast, but can get stuck

Does the greedy technique work for all cases? Not really.

Why not? That seems a simple algorithm, and we are only considering legal moves. The problem is it can't deal with a **local minima**. What is a local minima? It is a place on the map where the robot would have to go backward to find a good path. The easiest type of minima to visualize is a U-shaped area where the robot can get in but not back out. The Greedy Best-First algorithm is also not trying to find the shortest path, just a valid path:

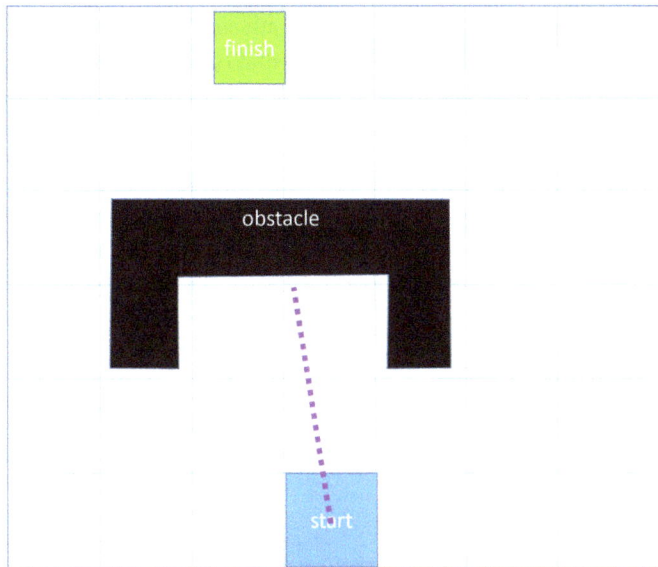

Figure 8.10 – A "local minima" can occur when no straight path exists,
and the robot will have to back up or reverse direction

If we want to find the shortest path, we need to do some more math.

A more systematic and mathematical way to approach finding the shortest path around obstacles for a grid search problem is the **A* algorithm**, first developed for Shakey the Robot.

Introducing the A* algorithm

Honestly, you can't really write a book about robotics without mentioning the A* algorithm. A* has its origins with *Shakey the Robot* at Stanford University back in 1968. This was one of the first map-navigating robots. Nils Nilsson and his team were trying to find a method to navigate Shakey around the hallways at Stanford and started trying different algorithms. The first was called *A1*, the second *A2*, and so forth. After several iterations, the team decided that a combination of techniques worked best. In computer science, A* means the letter A followed by anything else, and thus the A-star was named.

The concept of the A-star process is very much like what we have already been doing with our other path planners. Like the wavefront planner, we start by considering the neighbors around our starting location. We will compute an estimate for each square based on two factors: the distance from the starting location and the distance in a straight line to the goal. We are going to use these factors to find the path with the lowest cumulative cost. We calculate that cost by adding up the value for each grid square that is part of the path. The formula is as follows:

$$F(n) = g(n) + h(n)$$

Here, *F(n)* refers to the contribution of this square to the path cost, *g(n)* represents the distance from this square from the start position along the path chosen (that is, the sum of the path cost), and *h(n)* is the straight line distance from this square to the goal, which is a heuristic or estimate of the distance remaining to the goal. Since we don't know what other obstacles we have to go around later, we use this guess as a measuring stick to compare paths.

This value represents the cost or contribution of this square if it were a part of the final path. We will select the square to be part of the path that has the lowest combined cost. As with the wavefront planner, we keep track of the predecessor square or the square that was traversed before this one to reconstruct our path:

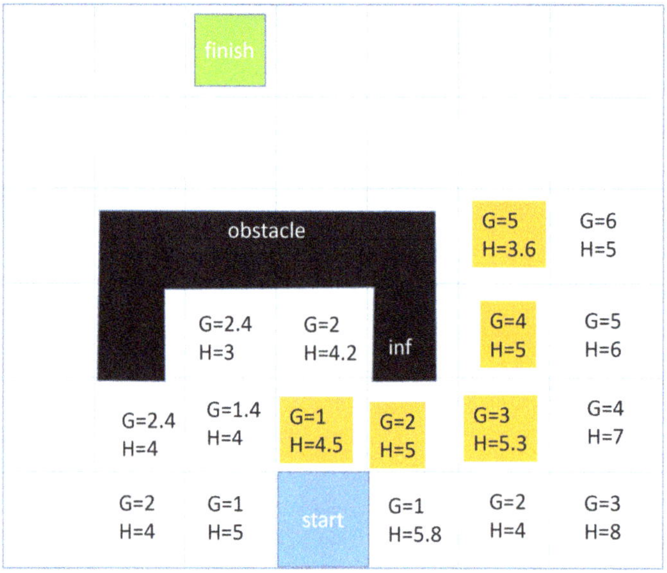

Figure 8.11 – The A-star computation uses the distance to start (G) and the distance to the goal (H)

The preceding diagram illustrates the A* algorithm. Each square is evaluated based on the sum of the distance along a path back to the start (*G*), and an estimate of the remaining distance to the goal (*H*). The yellow squares represent the path selected so far.

Let's illustrate how the A* algorithm works:

1. We keep a set of all the grid squares on the map we have computed values for. We'll call this `exploredMap`. Our map grid square object looks like this:

    ```
    # globals
    mapLength = 1280
    ```

```
mapWidth = 1200
mapSize = mapLength*mapWidth
map = []
```

2. Now, we will fill in our map with zeros to initialize everything. We will define the mapGridSquare function later in the code – it creates our data structures:

```
for ii in range(0, mapWidth):
    for jj in range(0,mapLength):
        mapSq = mapGridSquare()#defined later
        mapSq.position = [ii,jj]
        mapSq.sType =EMPTY
```

3. The next section creates all of the obstacles on the map. We put the location of which grid squares to *fill-in* or make impassable:

```
# create obstacles
obstacles = [[1,1],[1,2],[1,3],[45,18],[32,15]
.....[1000,233]]
# iterate through obstacles and mark on the map
for pos in obstacles:
    map[pos]. sType = OBSTACLE
pathGrid = []
```

4. Now, we declare our starting and ending positions:

```
START = [322, 128]
GOAL = [938,523]
exploredMap = []
A_Star_navigation(start, goal, exploredMap, map)
```

5. In this section, we are creating our data structures to keep track of all of the computations we make. The G value is the computed distance from the start, and the H value is the estimated distance to the goal. F is just the sum of these two. We also create a function to compute these values:

```
def mapGridSquare():
    def __init__(self):
        self.F_value = 0.0   #total of G and H
        self.G_value = 0.0   # distance to start
        self.H_value = 0.0   # distance to goal
        self.position=[0,0]    # grid location x and y
```

```
        self. predecessor =None    # pointer to previous
square
        self.sType = PATH
    def compute(self, goal, start):
        self.G_value = distance(goal.position,self.
position)
        self.H_value = distance(start.position,self.
position
        self.F_value = self.G_value + self.H_value
        return self.F_value
```

6. We need a function to trace the path from the goal back to the start once we've completed the map computations. This function is called reconstructPath:

```
def reconstructPath(current):
    totalPath=[current]
    done=False
    while not done:
        a_square = current.predecessor
        if a_square == None:  # at start position?
            done = True
        totalPath.append(a_square)
        current = a_square
    return totalPath
```

7. We create a findMin function to locate the grid block that we have explored with the lowest F score:

```
def findMin(map):
    minmap = []
    for square in map:
        if minmap == []:
            minmap = square
            continue
        if square.F_value < minmap.F_value:
            minmap = square
    return minmap
```

8. Then, we create the navigation function itself:

```
def A_Star_navigation(start, goal, exploredMap, map):
    while len(exploredMap>0):
        current = findMin(exploredMap)
        if current.position == goal.position:
            # we are done - we are at the goal
            return reconstructPath(current)
        neighbors = getNeighbors(current)
```

9. The neighbors function returns all the neighbors of the current square that are not marked as obstacles:

```
for a_square in neighbors:
    if a_square.predecessor == None:
```

10. We only compute each grid square once:

```
old_score = a_square.F_value


score = a_square.compute(GOAL, START)
```

11. Now, we look for the square that has the lowest G value – that is, the one closest to the start:

```
if a_square.G_value < current.G_value:
    a_square.predecessor = current
    current = a_square
    current.compute(GOAL, START)
    exploredMap.append(current)
```

So, in this section, we've covered the A* approach to finding the shortest path on a map, given that we know where all of the obstacles are in advance. But what if we don't? Another method we can use is the D* algorithm.

Introducing the D* (D-star or dynamic A*) algorithm

Earlier in the chapter, I talked about *a priori* knowledge. The A-star algorithm, for all its usefulness, requires that obstacles in the entire map be known in advance. What do we do if we are planning a movement into an unknown space, where we will create the map as we go along? If we have a robot with sensors, such as sonar or lidar, then the robot will be detecting and identifying obstacles as it goes. So, it must continually replan its route based on increasing information.

The A* process is only run one time to plan a route for a robot before it begins to move. **D***, a dynamic replanning process, is constantly updating the robot's path as new information becomes available.

The D* algorithm allows for replanning by adding some additional information to each grid square. You will remember that in A*, we had the G value (distance to the start along the path), and the H value (straight-line distance to the goal). D-star adds a tag to the square that can have several possible values:

- The square's tag could be NEW for a new square that had never been explored before.

- It could be OPEN for tags that have been evaluated and are being considered as part of the path.

- CLOSED is for squares that have been dropped from consideration.

- The next two tags are RAISED and LOWERED. The RAISED flag is set if a sensor reading or additional information caused the cost of that square to increase, and LOWERED is the opposite. For LOWERED squares, we need to propagate the new path cost to the neighbors of the now lower-cost square, so that they can be re-evaluated. This may cause tags to change on the neighboring squares. RAISED squares have increased cost, and so may be dropped from the path, and LOWERED squares have reduced cost and may be added into the path.

> **Note**
> Keep in mind that changes in cost values ripple through the D* evaluation of paths like a wave as the path is backtracked all the way to the start when the values change.

Another major difference between D* and A* is that D* starts at the goal and works backward toward the start. This allows D* to know the exact cost to the target – it is using the actual path distance to the goal from the current position and not a heuristic or estimate of the distance to go, as A* did.

This is a good time to remind you that all these grid-searching techniques we just covered are still variations of decision trees. We are going from leaf to leaf – which we have been calling grid squares, but they are still leaves of a decision tree. We set some criteria for choosing which of several paths to take, which make branching paths. We are working toward some goal or endpoint in each case. I bring this up because, in the next section, we will combine decision trees and the type of path planning we learned from the A* and D* algorithms to find a path through streets with a GPS.

GPS path finding

I wanted to have the opportunity (since we have come this far) to talk just for a little bit about **topological path planners**. This is an alternative method to the grid-based techniques we used in the preceding sections. There are types of problems and types of navigation where a grid-based approach is not appropriate or would require astronomical amounts of detailed data that may not be available or practical in a small robot.

As an example, I wanted to talk about how your GPS in your car finds a route along streets to reach your destination. You must have wondered about how that box has enough information in its tiny brain to provide turn-by-turn directions from one place to another. You may have imagined, if you stopped to think about it, that the GPS was using the same map you were viewing on the LCD screen to determine where you need to go. You would also think that some sort of grid-based search took place, such as the A* algorithm we discussed in such detail. And you would be wrong.

The data that the GPS uses to plan a route does not look like a map at all. Instead, it is a **topological network** that shows how streets are interconnected. In format, it looks more like a database of vectors (which have a direction and a magnitude, or distance), rather than an *X, Y* gridded raster map made up of pixels. The database format also takes up a lot less room in the GPS internal storage. The streets are divided by **nodes** or points where roads intersect or change. Each node shows which streets are connected. The nodes are connected by **links**, which allow you to traverse the data from node to node. The links represent the roads and have a length, along with cost data about the quality of the road. The cost data is used to compute the desirability of the route. A limited access highway with a high-speed limit would have a low cost, and a small side street or dirt road with a lot of stop signs would have a high cost since that link is both less desirable and slower.

The technique that most GPS path planners use is called **Dijkstra's algorithm**, after Edsger W. Dijkstra, from the Netherlands. He wanted to find the shortest path from Rotterdam to Groningen, back in 1956. His graph-based solution has withstood the test of time and is very commonly used for GPS routing. It's not of any help to us for our robot, so you can research this on your own.

We use the same procedures with the GPS road network database as we would when working the A-star process on a grid map. We evaluate each node, and progress outward from our start node, choosing the path that takes us closest in the direction of our destination:

Road Network Database has a list of nodes (circles) and links (lines).

Each node has a list of which links intersect at the node

Each link has a length and a cost (type of road), as well as a pointer to the node at each end

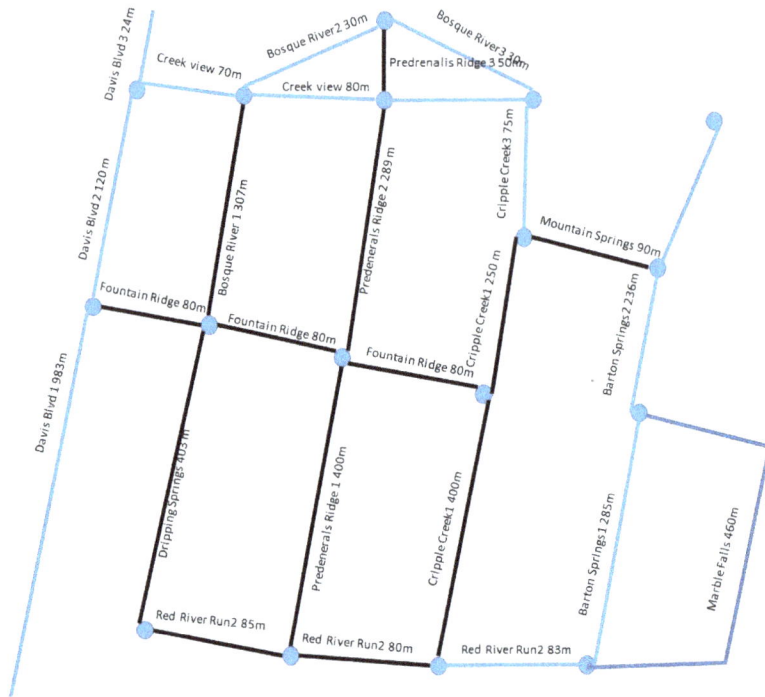

Davis Blvd 3 24 m
Creek view 70m
Bosque River2 30m
Bosque River3 30m
Predrenalis Ridge 3 50m
Creek view 80m
Davis Blvd 2 120 m
Bosque River 1 307m
Predenerals Ridge 2 289 m
Cripple Creek3 75m
Mountain Springs 90m
Fountain Ridge 80m
Fountain Ridge 80m
Fountain Ridge 80m
Cripple Creek1 250 m
Barton Springs2 236m
Davis Blvd 1 983m
Dripping Springs 403 m
Predenerals Ridge 1 400m
Cripple Creek1 400m
Barton Springs1 285m
Marble Falls 460m
Red River Run2 85m
Red River Run2 80m
Red River Run2 83m

Figure 8.12 – A road-based network can be represented as a series of nodes (circles) and links (lines)

Many GPS systems also simultaneously try to backward-chain from the endpoint – the goal or destination – and try to meet somewhere in the middle. An amazing amount of work has gone into making our current crop of GPS systems small, lightweight, and reliable. Of course, they are dependent on up-to-date information in the database.

Summary

Well, this has been a very busy chapter. We covered the uses of decision trees for a variety of applications. The basic decision tree has leaves (nodes) and links, or branches, that each represent a decision or a change in a path. We learned about fishbone diagrams and root cause analysis, a special type of decision tree. We showed a method using `scikit-learn` to have the computer build a classification decision tree for us and create a usable graph. We discussed the concept of random forests, which are just an evolved form of using groups of decision trees to perform prediction or regression. Then, we got into graph search algorithms and path planners, spending some time on the A* (or A-star) algorithm, which is widely used for making routes and paths. For times when we do not have a map created in advance, the D* (or dynamic A-star) process can use dynamic replanning to continually adjust the robot's path to reach its goal. Finally, we introduced topological graph path planning and discussed how GPS systems find a route for you to the coffee shop.

In our next chapter, we'll be talking about giving your robot an artificial personality, by simulating emotions using a Monte Carlo model.

Questions

1. What are the three ways to traverse a decision tree?

2. In the fishbone diagram example, how does one go about pruning the branches of the decision tree?

3. What is the role of the Gini evaluator in creating a classification?

4. In the toy classifier example using Gini indexing, which attributes of the toy were not used by the decision tree? Why not?

5. Which color for the toys was used as a criterion by one of the classification techniques we tried?

6. Give an example of label encoding and one-hot encoding for menu items at a restaurant.

7. In the A* algorithm, discuss the different ways that G() and H() are computed.

8. In the A* algorithm, why is H() considered a heuristic and G() is not? Also, in the D* algorithm, heuristics are not used. Why not?

9. In the D* algorithm, why is there a RAISED and LOWERED tag and not just a CHANGED flag?

Further reading

- *Introduction to the A* Algorithm*: https://www.redblobgames.com/pathfinding/a-star/introduction.html

- *Introduction to AI Robotics* by Robin R. Murphy, MIT Press, 2000

- *How Decision Tree Algorithm Works*: https://dataaspirant.com/2017/01/30/how-decision-tree-algorithm-works/

- *Game Programming Heuristics*: http://theory.stanford.edu/~amitp/GameProgramming/Heuristics.html

- *D*Lite Algorithm Blog (Project Fast Replanning)* by Sven Koening: http://idm-lab.org/project-a.html

- *Graph-Based Path Planning for Mobile Robots*, Dissertation by David Wooden, School of Electrical and Computer Engineering, Georgia Institute of Technology, December 2006

- *The Focused D* Algorithm for Real-Time Replanning* by Anthony Stentz: https://robotics.caltech.edu/~jwb/courses/ME132/handouts/Dstar_ijcai95.pdf

9

Giving the Robot an Artificial Personality

When a person thinks of a robot with AI, what many consider AI is a robot that has emotions, feelings, a state of mind, and some sort of model or concept of how humans think or feel. We can call this form of AI an **artificial personality**. While giving a robot feelings is definitely way beyond the scope of this book (or current technology), what we can do is create a simulation of a personality for the robot using standard computer modeling techniques, such as Monte Carlo analysis, and finite state machines.

In this chapter, we will cover the following topics:

- What is an artificial personality?
- A brief introduction to the (obsolete) Turing test, chatbots, and **generative AI** (**GenAI**)
- The art and science of simulation
- An emotion state machine
- Playing the emotion game
- Creating a model of human behavior
- Developing the robot emotion engine

Technical requirements

We will not be introducing any new programming libraries in this chapter. We will be building on the voice system we constructed previously. All you will need is imagination and some writing skills.

You'll find the code for this chapter at `https://github.com/PacktPublishing/Artificial-Intelligence-for-Robotics-2e`.

What is an artificial personality?

Hollywood and the movie industry have turned out some very memorable robots. You can think of R2D2 and C3PO, the Laurel and Hardy of science fiction. What do you like most about these two? Could it be their personalities? Think about this a bit. Even though R2D2 is mostly a wastebasket shape with a dome head and no face, he has a definite personality. You'd describe him as *feisty* or *stubborn*. The *Robots and Androids* website (`http://www.robots-and-androids.com/R2D2.html`) described him in this way:

> *The droid is shown as being highly courageous, unwavering in loyalty, and a little bit stubborn. He never backs down from a mission, even when the odds appear to be stacked against him. His personality is often contrasted against that of [C3PO], who is fussy and timid.*

This is pretty impressive for a robot who never says a word and communicates with beeps and whistles.

What were other movie robots that made an impression on you? Certainly, WALL-E, the lovable trash-compacting robot from the eponymous movie, is a favorite. WALL-E also had a small vocabulary, consisting of only his name, much like a Pokemon. WALL-E displayed a lot of emotion and even developed hobbies, collecting and repairing old trash. You may also remember M-O, the tiny, obsessive cleaning robot that gets frustrated with all of the dirt WALL-E brings in.

So, one thing that we might do as robotics creators and designers is to imbue our robot with some sort of personality. This has the advantage of letting humans relate to the robot better. It also gives the false impression that the robot is much smarter and is capable of more than it really is. This does give the advantage of the robot being more engaging and interesting.

We can also infer from the examples given to us by R2D2 and WALL-E that less can be more when it comes to communication – we need to not just have words but also body language and sound.

What we are going to do in this section is develop an artificial personality for our robots. While it is impossible for us to give a robot actual emotions or feelings, however you might define that, we can create a simulation of personality that will provide a convincing illusion. I think this is a meaningful exercise because the current state of the art in robotics demands some sort of personality and consistent demeanor from robotics that interact with humans.

The cartoon seen in *Figure 9.1* expresses how people and robots might interact. The robot has had enough with picking up toys and wishes to show its feelings:

Figure 9.1 – A person and a robot interacting

There is a lot of work going on in this area right now with digital personal assistance, such as Apple's Siri and Amazon's Alexa. Note that these robots, or AIs, have distinct names and voices, but I feel that they are very similar in personality and capability. There is some distinction when you ask Siri or Alexa a personal question, such as how old they are (Siri is considerably older than Alexa).

In this chapter, we are going to take tools from the science of simulation – specifically, **state machines** and **Monte Carlo analysis** – and use them to form a model of a personality for Albert, the robot. We already have a fairly powerful tool, in the Mycroft speech system we used to tell knock-knock jokes. We will be extending Mycroft with some new skills and capabilities, as well as developing cybernetic emotions, both for our robot and for the robot's opinion of what we, the humans, are feeling.

I want to emphasize that we are simulating emotions, not creating an emotional robot. Our simulation bears the same resemblance to real emotions as a flight simulator bears to the space shuttle – both provide the same information, but the space shuttle flies around the Earth in 90 minutes, and the flight simulator never moves.

Let's talk about what we are trying to accomplish in this section by first talking about the famous Turing test – can we create a robot that interacts in a way that's indistinguishable from a human being?

A brief introduction to the (obsolete) Turing test, chatbots, and generative AI

Alan Turing proposed his famous test, which he called *The Imitation Game*, in a paper titled *Computing Machinery and Intelligence*, published in 1950 in the journal *Mind – A Quarterly Review of Psychology and Philosophy* (see `https://www.abelard.org/turpap/turpap.php#the_imitation_game`). In the original text, Turing imagined a game where a player would have to guess the sex – male or female – of a hidden person by typing questions on a teletype. Then, he suggested that a truly intelligent machine would be one where you would not be able to distinguish if the hidden personality on the other end of the teletype was a human or a computer software program.

> **Note**
>
> The movie *The Imitation Game* stars Benedict Cumberbatch as Alan Turing and features his role in breaking German code in WWII as part of the mathematicians of Bletchley Park. The title refers to Turing's original name of the famous test that bears his name.

These days, you may talk to computer software many times a day and not realize that you are not speaking to a human. Robocalls and chatbots may call you on the telephone or engage you in a fake political conversation on Twitter. The Turing test has been won by the machines (`https://www.nature.com/articles/d41586-023-02361-7`), but have we developed intelligence in computers? Not at all – we have just become very clever at simulating conversation. Recently, robotics experts have suggested replacing the Turing test with a more demanding and difficult assessment of a computer's cognitive skills and self-understanding: `https://techxplore.com/news/2023-11-redefining-quest-artificial-intelligence-turing.html`.

We are going to use another tool that Alan Turing mentioned in his paper – **state machines**. We will use state machines to define and model the emotions of our robot in the *An emotion state machine* section.

Let's now go back to the concept of a **chatbot**. A working definition may be a software program designed to engage a human in conversation or to interact with a person via voice or text. While most chatbots are up-front about being computer generated, there are a lot of chatbots that are not – including Twitter chatbots that seek to influence elections or public opinion. Many chatbots are gainfully employed answering tech support phone calls, making marketing calls, and entertaining users as the dialog of **non-player characters** (**NPCs**) in games.

According to the article *Ultimate Guide to Leveraging NLP and Machine Learning for Your Chatbot* by Stefan Kojouharov, published by `chatbotslife.com` in 2016, chatbots come in two flavors:

- **Retrieval-based**: These chatbots rely on stored phrases and words, and use the software decision-making to select which reply is most appropriate. There may be some keyword recognition and noun-subject insertion involved, but the main action is to select the most appropriate phrase.

- **Generative-based**: These chatbots make up new sentences based on **parts of speech** (**POS**) and the robot's divination of your intent. They can be thought of as machine translation engines that translate the input (your text or speech) into an output (the robot's reply). As you might imagine, the generative-type chatbot is far more difficult to achieve, which is why we will be using a retrieval-based approach. Recent developments from projects such as ChatGPT have redefined what is possible for chatbots. The term **Generative Pre-trained Transformer** (**GPT**) refers to three characteristics of this **neural network** (**NN**):

  - **Generative**: The model is capable of creating new text phrases, rather than just repeating or categorizing text

  - **Pre-trained**: The model is pre-trained on enormous datasets – over a trillion examples – to understand language and the relationship between words

  - **Transformer**: The NN architecture uses transformers to process an entire sentence at a time, learning both the word meanings and relationships in a sentence (positions of words in a sentence)

In the simplest form, the GenAI NN predicts what word is most likely to come next in a sentence, based on training on billions of examples.

We will use GenAI to develop some of our text outputs but apply these to a classical chatbot, thus having the best of both worlds – not having to generate text ourselves, and having a system where we have control over inputs and outputs and can trust the system to control a robot.

There are two other details we need to attend to. Chatbots can be designed to handle either short conversations or long conversations. The vast majority of chatbots – and that includes digital assistants such as Siri, Alexa, and Mycroft – are designed for very short conversations. *Siri, what is the weather? There is a 20% chance of rain. The high is 88 degrees.* That's it – the whole conversation in three sentences and two interactions. If you ask another question, it starts a new conversation with little reference to the previous one.

A more difficult task is to have a longer conversation with several interactions and even a selection of topics. This requires the computer to keep track of context or what information has been discussed and might be referred to again.

We will be attempting to teach our robot to be able to have medium-length conversations at a 7-year-old level. I'll define medium length to be between two and six interactions.

Now, before we proceed with our robot, let's quickly discuss some statistical distributions because I've found the ability to use Monte Carlo analysis – and to create custom random number distributions – very useful in robotics.

The art and science of simulation

What is simulation? A **simulator** is a computer model of the physical world. You are probably familiar with flight simulators, which provide sensations and interactions of flight without leaving the ground. There are also a lot of other types of simulations and simulators. We could have a medical simulator that mimics diseases or responds to treatments. It could be a financial simulation that models profits on the stock market based on trends. There are structural simulations that model loads on bridges and buildings to see whether the materials are adequate.

The most common way of creating a simulation is by building a **physics model** of the item under test. For a flight simulator, this means plugging in formulas for the four forces on an airplane or a helicopter – lift, gravity, thrust, and drag. Each factor has parameters that affect its performance – for instance, the lift is a function of the speed through the air, the weight of the aircraft, the size of the wing, and the angle of attack, or the angle between the wing and the wind. Vary any of those, and the lift changes. If the amount of lift exceeds the force due to gravity (that is, the weight of the aircraft), then the aircraft flies. The simulation sets up a time-step interval, just like our control loop for the robot, and computes the forces on the aircraft for each time step. We can then apply controls and see how our model performs. Models just like this are used to predict performance in advance of building a prototype or test airplane.

Another type of simulation is called a **Monte Carlo model**. The Monte Carlo method uses **probability theory** to replace sophisticated physical models with a variation of random numbers that approximates the same result. If you wanted to create a computer model of flipping a coin, you wouldn't spend a lot of time determining the physical properties of a nickel or modeling the number of flips in the air based on force. You would just pick a random number from 1 to 100 and say the result is heads if the number drawn is less than 50 and tails if it is greater than 50. That, in essence, is the Monte Carlo method. There are a lot of physical processes that can be approximated and studied using this technique, where outcomes can be described in terms of probabilities.

We can apply Monte Carlo analysis to model people going through security at an airport. If you had a copy of a typical schedule for the airlines and the average number of passengers per flight, you would know the daily traffic at the airport. The difficult bit would be modeling when people would arrive for their flight. Let's imagine that we commissioned a study and determined roughly that 50% of people arrive 1 hour early, 25% arrive 2 hours early, and the rest are evenly distributed between 2.5 hours and 30 minutes, with 1 passenger out of every 200 missing their flight by being late. We can approximate the passenger arrival function with two standard distributions (bell curves) and two uniform distributions (boxes):

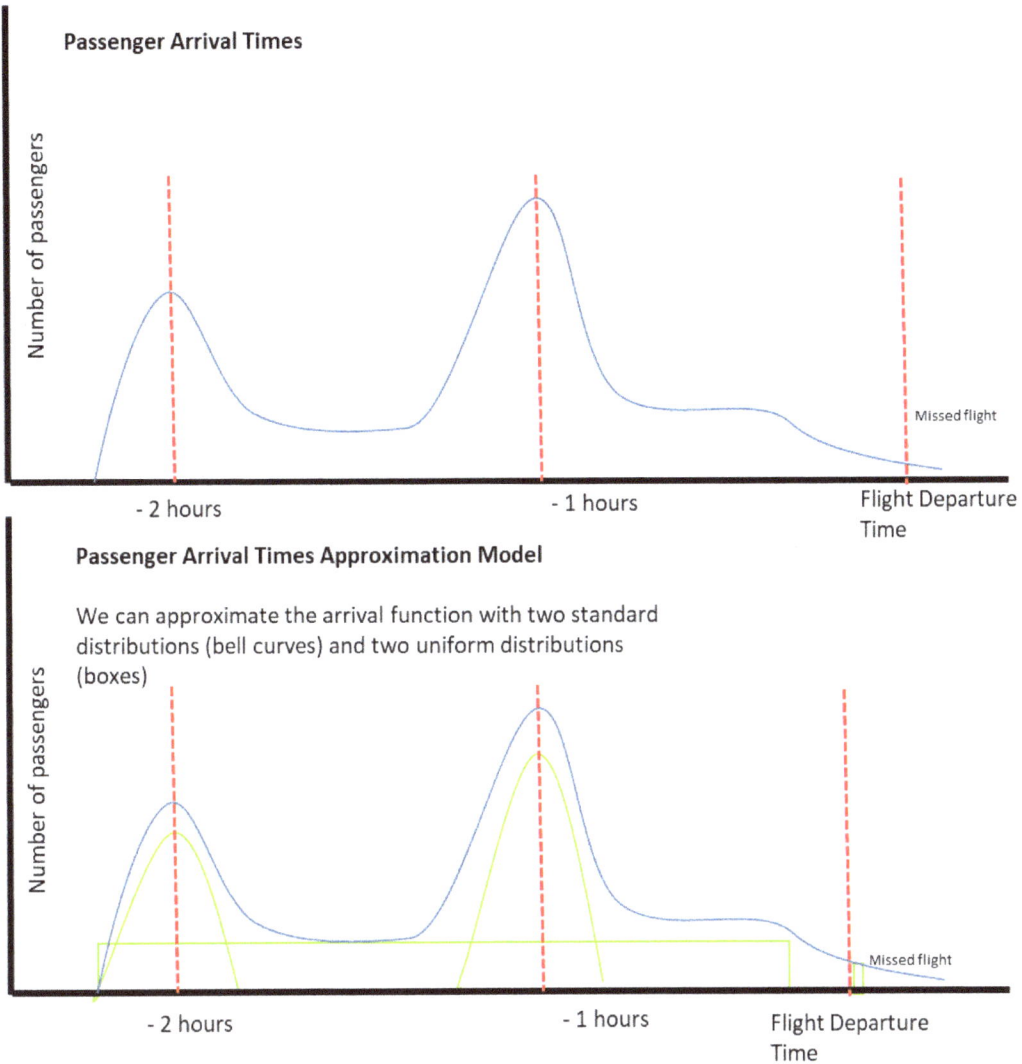

Figure 9.2 – A complex probability distribution function can be
approximated by the union of simpler distributions

This information allows us to create sample sizes of passenger arrivals and thus obtain an estimate of the length of lines at security. We would add some distribution of values of how long it took to get through security, and we would have our model. We have a list of flights, and for each flight, we generate a distribution of passenger arrival times based on the study by assigning random numbers to each passenger and separating them out into groups based on arrival time.

Let's assign random numbers from 1 to 100 to our 212 passengers arriving for the 08:00 flight from Dallas to Washington DC. Now, we assign an arrival time based on that number – if the number is from *1* to *25*, the passenger arrives 2 hours earlier than the flight (06:00). If the number is from *26* to *75* (the next 50%) then they arrive 1 hour early (07:00). The rest of the passengers, having numbers from *76* to *100*, are assigned random times between 2.5 hours early to 30 minutes early. And we pick 1 unlucky passenger out of our 212 to miss the flight completely by arriving late. Since there is some variability in people who intend to arrive exactly 2 hours ahead, but are delayed or advanced slightly, we can **dither** or add a small random distribution factor to each arrival time of plus or minus 10 minutes. We now have a statistically correct distribution of people arriving for a flight. Now, repeat this process for the other 1,849 flights leaving today from this airport. You can see from this model that we can still put a great deal of science into picking random numbers.

So, how do we make a **probability distribution function** (**PDF**) that looks like this (see *Figure 9.2* top graph)? The answer is we combine several distributions. In the lower graph of *Figure 9.2*, you see two standard distributions (the humps in green) and one uniform distribution (the lower green box) that combined make our overall correct function.

As you might have guessed from my example, the true secret of making this technique work is picking the correct distribution of random numbers. You may think, "Hey, random numbers are random numbers, right?" But this is not correct at all. You have heard of a **standard distribution** or the **bell-shaped curve**. Most of the values are in the center, with fewer and fewer as you go away from the center. Many natural processes follow this bell-shaped curve, including grades on a test, how tall people are, or how many grapes grow in a bunch. When you do Monte Carlo analysis, most often we use some form of normal or standard distribution.

Can you think of other types of random numbers? The second most common type of random number is a **uniform distribution**. Each number has the exact same probability of being selected, which makes for a flat distribution curve. Another name for this uniform distribution is **white noise**. Uniform distributions do occur in analysis, and if we want to add noise to an image, for example, we will use a uniform distribution. But in modeling emotions and people, normal or standard distributions are the rule.

You may find that in using Monte Carlo modeling, a standard distribution or uniform distribution just won't work. Then, you can create a custom distribution, as I did in the airport example, where we used re-sampling to change a uniform distribution to a custom distribution fitting our passenger arrival model.

Figure 9.3 shows the various shapes of distributions various kinds of random number generators produce, along with their commonly used names – the bell curve and the flat line. The lognormal graph looks to me like a ski slope – what do you think it looks like?

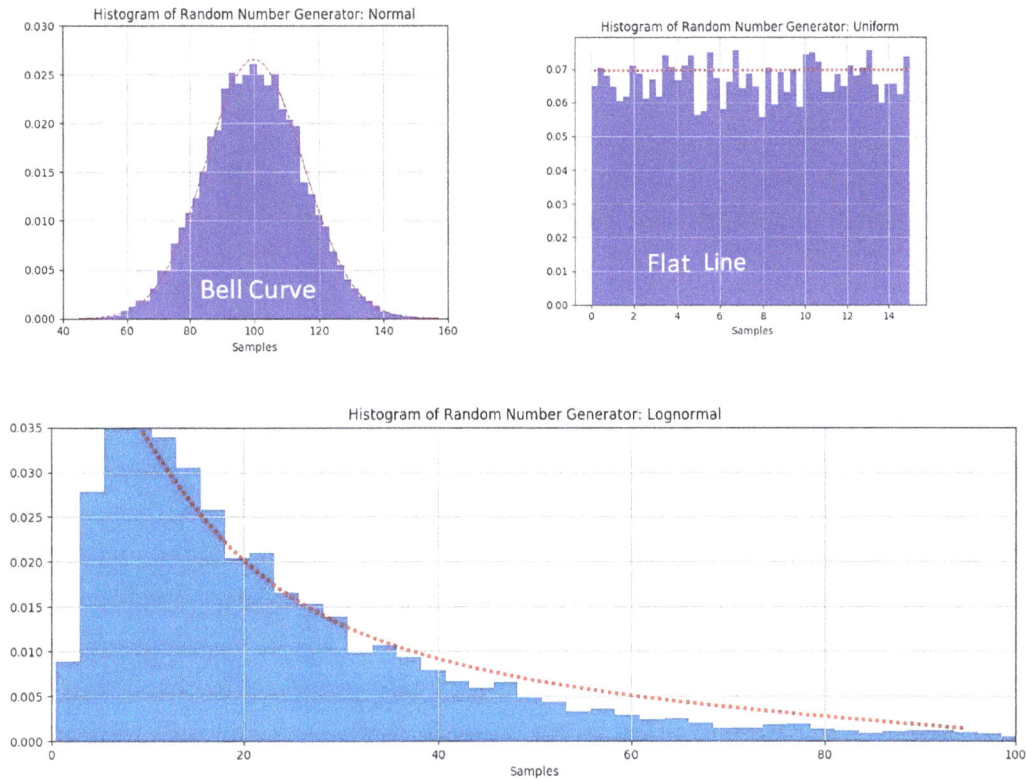

Figure 9.3 – Various types of random number generators and the distributions they produce

Our next task is to generate a personality simulation for our robot, and we will be using Monte Carlo modeling plus a state machine to create a model of our robot's emotions. Let's first discuss what an emotion state machine is.

An emotion state machine

What is a state machine? We covered state machines in the systems engineering section of *Chapter 2* – state machines are a technique for understanding or modeling automation or computer programs. A **state** is a set of conditions that exist at the present. I like to think of a state as being a set of functions that are constrained by limits. The machine (our robot) changes from state to state based on some event that causes the state to change.

Let's work through a quick refresher example. Let's take a slice of bread. When we get it, it is already baked and sliced, so its initial state is as a slice of bread. If we subject the bread to infrared radiation (that is, heat), then the surface becomes caramelized, and we call that toast. The state of the bread has changed, along with its taste and texture, from baked bread to toast. The event that caused that transition was the act of heating the bread in a toaster. This is pretty simple stuff, and I'm sure you have encountered state machines before.

Now, let's think about our model of robot emotions. We can start by listing what emotions we want our robot to have:

- Happy
- Welcoming
- Friendly
- Curious
- Positive
- Energetic

Then, we can list the opposites of those emotions:

- Sad
- Distant
- Unfriendly
- Frustrated
- Tired

These are the list of emotions I wanted to simulate in our robot. I looked at the different interactions the robot might have and how a human version of the robot would react:

Robot Emotion Model

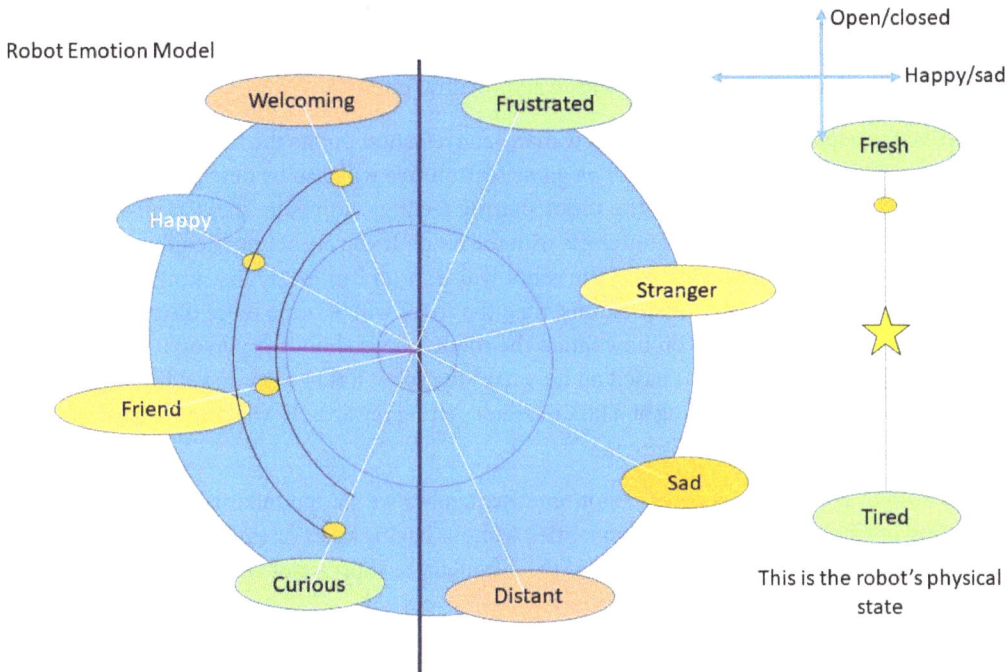

Figure 9.4 – Robot emotional state machine model

In the preceding diagram, we model the overall emotive state of the robot by plotting the various emotional components or states on a polar plot, as follows:

- Each emotional element has a vector or direction. On the left side of the plot are happy, friendly, open feelings, and on the right side are sad, closed, distant feelings.

- The distance from the center indicates the strength of the emotions.

- The yellow circles are the robot's current value on each axis.

- The purple line is the vector sum of those emotions, which gives us the overall mood of the robot. The arcs show the minimum and maximum magnitude of emotion.

We take the area made up of the four components and find the center of that area, and that is the overall state of the robot. In this example, the predominant emotion is *friendly*. Since the robot's physical state determines whether it feels *tired* or not, that data is computed separately – this lets the robot feel tired but friendly, or refreshed but sad.

The robot starts in a state of *happy*, *distant*, *curious*, and *fresh*. That is to say, it feels happy, it is not in an interaction, so there is no one to be friendly to, and it will be curious about its surroundings. As the environment changes, the robot's state will change. For example, we have about a 2-hour run time on Albert's current battery. It starts off in a *fresh* state and will get more and more tired as it approaches the 2-hour mark. We are going to use a simple timer to create this state, but you could use the voltage sensor in the robot arm to provide information about how fresh the robot's power supply is. Each emotion exists along an axis and all cross a central or neutral point.

We will drive the *happy-sad* axis based on how many conversation points the robot is earning. We'll cover this in detail in the *Playing the emotion game* section. We will also be describing an empathy function for our robot that simulates the robot sharing feelings with you. The *distant-welcoming* aspect is controlled by how the human interacts with the robot. If you are acting friendly to the robot, then it will be welcoming. If you are new, the robot will start off cautiously in asking questions or interacting. If you are not answering questions, then the robot will become more distant. Likewise, the *friend-stranger* aspect is based on how much the robot knows about the person it is talking to. Finally, the *curious-frustrated* axis is based on how hard the robot has to work to get information or to accomplish a task. If it keeps failing at a task or is not getting rewards, it will become frustrated in its expressions and vocabulary.

How does a robot with no face express emotions? Remember we started talking about Hollywood robots, many of whom have distinct personalities without having faces or even heads. We will use body language with the robot arm and changes in vocabulary to express our simulation of emotions. And we will have the robot be open about how it is feeling and why.

Our goal for this part of the robot development is to provide a human interaction framework that invites people to talk to the robot and feel welcome or wanted. I'm modeling this interaction from watching my grandchildren. I wanted a robot that they would enjoy and find interesting. What we want to do is develop the ability for the robot to convey intent, or to provide a simulation that the robot has needs and desires. We are going to do this by creating a game that the robot will play with humans – let's call this the emotion game.

Playing the emotion game

So, what exactly is this game? What we want the robot to do is ask questions of another person and to use conversation to get personal information out of them so that the robot then can use that information in replies. In this game, the robot will be trying to score points by getting the human to interact socially. The robot will gain points by getting information from the person so that it can *get to know them better*. The trick for us is we need the robot to keep this information and remember it. We will be saving all this information and using it to modify the actual code the robot is running, thus implementing **machine learning** (ML) into the conversation. The robot will be using the same type of point-reward system we used in training the robot arm back in a previous chapter. Once the robot has scored by learning a fact, it will no longer get rewarded for that fact and will go on to get other facts. Once it has learned all the facts, it will end the conversation. In practice, I think most users will want fairly short conversations, so we are setting our goal to have between two and six interactions per conversation.

Let's quickly summarize how the game will be played:

1. The user will always initiate a conversation by saying the robot's wake phrase, which right now is *Hey, Albert*.

2. The robot will respond using the Mycroft speech engine with a beep.

3. The user will initiate a conversation using a wake word, which is some version of *Hello, robot*.

4. The robot will then try to earn rewards by getting points, which it does by asking questions.

5. The emotion engine will respond to the human answers by modifying the robot's emotions.

6. We will rank questions by importance – let's say the robot gets 10 points for learning your name, 9 points for learning your age, and so on.

7. Once a fact is learned, no more points are earned, so the robot won't repeat the questions.

The facts we want the robot to know about each person are the following:

- Your name
- Your age
- How are you feeling today?
- What is your favorite food?
- What is your favorite book?
- Do you go to school?
- If so, what is your favorite subject?
- When is your birthday?
- What is your favorite song?
- Do you like the color pink/singing/dancing/dinosaurs/race cars/building things/robots/airplanes/spaceships?
- Do you brush your teeth?
- Do you like knock-knock jokes?

As part of the learning game for the robot, we will adjust the robot's emotions as it learns and interacts, by adjusting the levels of the eight emotions (or four emotion types) we provided to the robot. We will particularly pay attention to the place where the eight emotions balance – are they on the *happy/ friendly/curious* side of the graph, or more on the *sad/frustrated/distant* side?

Since we are trying to get some personal information about people talking to the robot, they may want to find out more about the robot too. So, to respond, our robot will also have a backstory or a biography that it will use to answer questions about itself. We'll give the robot a little narrative:

- His name is Albert.

- His full name is Albert Robot the Second.

- He is 8 months old.

- He was made by Grandad.

- He was born on January 28, 2023.

- He likes the color green.

- His favorite food is electricity.

- His favorite author is Isaac Asimov.

- He does not go to school but loves to learn.

- His job, his hobby, and his passion is picking up toys.

- If you ask him how he feels, he will tell you which emotional state is highest, plus how fresh or tired his battery is. We want him to occasionally interject how he is feeling into the conversation without being asked.

> **Note**
>
> I've been addressing this robot as *he* and *him* all through the book. This is just an anthropomorphic projection on my part and is implying features that a robot just does not have. The primary reason for Albert's identity is his voice – I used a male voice for his synthesizer, mostly because I wanted it to stand out from all the female GPS and personal assistant computer voices around. Please feel free to create any voice you like – there are a lot of female voices available, and you can create whatever persona you like in your robot and give them any form of address. It is interesting that we tend to do this even with non-humanoid robots.
>
> According to Colin Angle, CEO of iRobot, over 80% of Roomba owners have given their robot a name, including him (`https://slate.com/technology/2014/03/roomba-vacuum-cleaners-have-names-irobot-ceo-on-peoples-ties-to-robots.html`). You don't see people naming their toaster ovens or stand mixers, no matter how attached they are to them. Albert the robot may very well be getting a sex-change operation, or get a sister, as my granddaughter gets a bit older.

We will also be using the emotional state to set the robot's body language, which is primarily how he is carrying his robot arm. If he is happy, the arm will be extended with the hand pointed upward. If he is sad, the arm will be close to his body and his hand will point down. We will store all this information to give the robot a consistent set of answers to personal questions.

Creating a model of human behavior

For the robot to support conducting a conversation, we must also have a model of how the human it is talking to is feeling. You may have had a friend or relation who went on talking about themselves and were oblivious to how you were feeling or reacting to their conversation. We don't want that type of robot personality. Therefore, the robot has to have some internal representation of how it thinks you are feeling. We will not be using vision for this function, so the only way the robot knows how you are doing is by asking questions and evaluating the use of language.

We will give the robot a human model similar to our state machine but with just four emotions on two axes: *happy/sad* and *friendly/distant*. The robot will assume that everyone is somewhere in the middle when the conversation begins. The robot can use clues in language to understand how you might be feeling, so we will assign *colors* or shades of emotions to words to help drive the robot's human model. This can be illustrated by the following diagram:

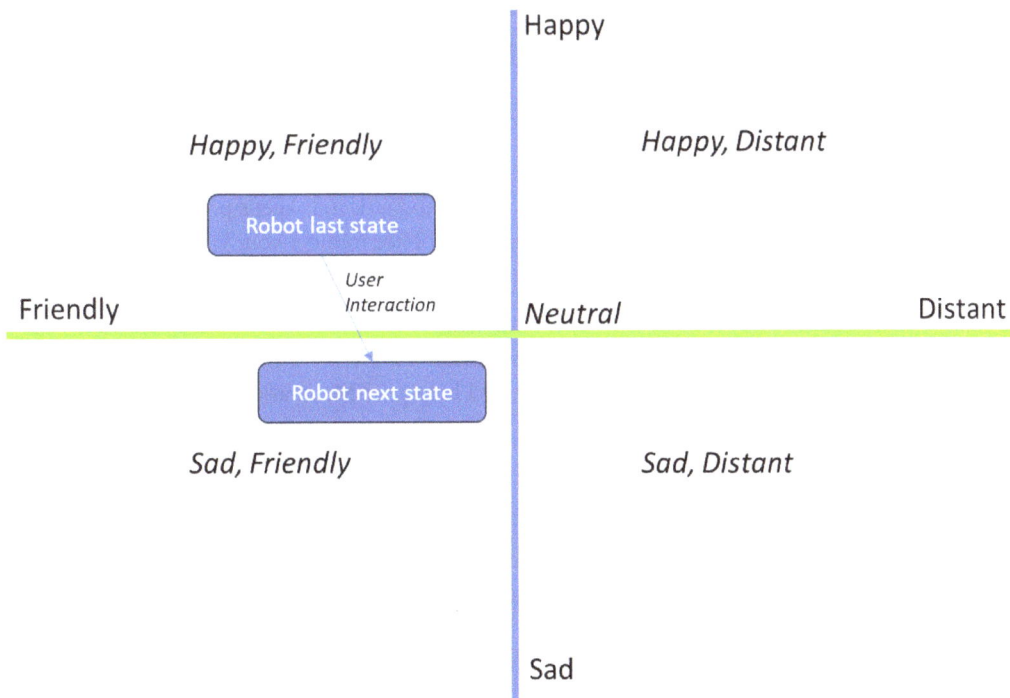

Figure 9.5 – A simplified emotional continuum

Different emotions from the person will drive different responses from the robot. We particularly want the robot to pay attention to clues that the human is becoming frustrated due to the robot not understanding or responding in the way the person wants. This brings us to actually selecting and designing an artificial personality for our robot.

Integrating an artificial personality into our robot

We need to set some guidelines and parameters that will determine what kind of personality the robot has. We can start by listing some types of personalities that a robot might have and what kind of role they might be playing. This is particularly important because this robot will primarily be interacting with children. Let's try a few out and see what fits:

- **Teacher/professor**: The robot is attempting to convey information and to teach or cause the person to change behavior in some way. The robot presents itself as an authority and provides facts and information about itself, such as "I'm a robot. Do you know what a robot is?"

- **Hyper-friendly**: The robot is playful and excited about talking with humans, aggressively engages people in conversation, and asks lots of questions. The robot expresses a lot of enthusiasm and encouragement. "I love my job. Don't you love your job? It's so great! I just love talking to people! Do you want to hear a joke?"

- **Chummy and friendly**: The robot is trying to make friends and be helpful. "Hi! Nice to see you! What is your name?"

- **Friendly but neutral**: The robot is not quick to share information but does want to talk to you. "Hello, I am Albert. How are you today?"

- **Feisty**: The robot is supportive, friendly, and wants interaction. It keeps prompting for more information. It will repeat questions if necessary. "Hi! Nice to meet you. What is your name? How old are you?"

- **Somber and staid**: The robot is stuffy and authoritative. Think of having a conversation with a British butler. The robot uses formal English. (This is apparently the default non-personality of many digital assistants.) "Greetings. How may I help you? May I ask you your name?"

- **Very neutral or robotic**: The robot expresses no opinions and gives no information about itself. It expresses no emotions or interjections. "Hello. I am a robot. State a command."

- **Annoyed and grumpy**: The robot is not very happy about having to pick up toys and does not mind saying so clearly and distinctly. This personality is aiming for comedy, not anger. "So, you are back again. I guess this means more toys to pick up."

We can see from this list the sort of gamut of personalities we can choose for simulation. I'm going to choose the *chummy and friendly* type of personality since that is what I feel will go over best with my grandchildren. You, of course, may choose any of these you please or add some more. Now, let's see how we can proceed with this.

Constructing a personality

In a quick review, what parts do we have so far for our robot with an artificial personality?

- We have simulation tools to model emotions.

- We have a state machine with six types of emotions that we can mix together.

- We have a backstory or biography of the robot's personal information.

- We have picked a personality to emulate.

- We have the concept of a game engine where the robot is trying to score points for collecting information by engaging in small talk or **phatic communication**.

- The robot will change emotions based on conversations. Emotions will be expressed either by asking the robot how he is feeling or by the body language of the robot's arm.

Now, we need some sort of framework to put all of these pieces together that can roughly carry on a conversation we can adjust and manipulate. For that, we are going to revisit an old, old friend. Let's talk, once again, about **ELIZA**.

ELIZA is probably the original conversation engine software program. It was invented back in 1964 by Joseph Weizenbaum, professor at MIT, a Jewish German refugee from WWII, and considered one of the fathers of AI. ELIZA was named after Eliza Doolittle, the main character in George Bernard Shaw's book *Pygmalion*. In Greek mythology, Pygmalion was a sculptor who fell in love with one of his statues. He prayed to Aphrodite, who granted his wish for the statue to come to life. This is very apt for what we are doing – bringing our robot to simulated life via this weird chain of events, titles, and Greek mythology.

ELIZA was a simulation of a Rogerian therapist. What is a Rogerian therapist? According to *Psychology Today* (`https://www.psychologytoday.com/us/blog/what-doesnt-kill-us/201210/unconditional-positive-regard`), Carl Rogers (1902-1987) is the founder of **client-centered therapy**, which believes in **unconditional positive regard** (**UPR**). This is to say that the therapist keeps their own opinions out of therapy sessions and primarily mirrors the patient's words back at them. If the patient says, "I've always hated cats," the therapist may respond, "Why do you think you hate cats?" rather than saying, "I have a cat and they are lovable creatures." So, ELIZA, the computer program, reflects back at you what you are saying without interjecting anything else. You have a conversation with ELIZA via what we now call *texting* and what we used to call *typing*. Here is an actual ELIZA interaction between me and an open source version of ELIZA by Wade Brainerd, which can be found at `https://github.com/wadetb/eliza`. The user is typing at the > prompt. By the way, I love cats – this is just an example. Please don't tell my cat about this example:

```
ELIZA: How do you do. Please tell me your problem.
> I hate cats.
You say you hate cats?
```

```
> Yes.
You seem to be quite positive.
> I hate cats.
Can you elaborate on that?
> I don't like cats because they smell bad.
Don't you really like cats because they smell bad?
> No. Don't like cats.
Are you saying no just to be negative?
> No.
You are being a bit negative.
> I am
Is it because you are that you came to me?
> No
Why not?
> I am afraid of cats
How long have you been afraid of cats?
> About a year
I'm not sure I understand you fully.
> 1 year
Please go on.
> I have been afraid of cats since the cat jumped on my head.
Your head ?
> yes
```

So, you can see that ELIZA repeats back phrases to the user, reflecting what the user is talking about. When I said, "I am afraid of cats," the program replied, "How long have you been afraid of cats?"

ELIZA goes through several steps in recognizing and responding to sentences. Most of the program is not hardcoded, and it works from a series of scripts that you can edit and replace to basically reprogram the personality of the program – which is why we are talking about ELIZA right now:

1. The first step is to divide the sentence into words separated by spaces. The program also replaces all capital letters with lowercase.

2. The program replaces words with several versions with a standard word. For example, the program replaces *cant* with *can't* in case you left out the apostrophe. This is called **preprocessing**.

3. The program looks for **keywords** that it knows about. A simple keyword is *sorry*. Any sentence with *sorry* gets a response such as *please don't apologize*. Keywords are collected and prioritized by the order they appear in the script file.

4. The program looks for *decomposition* patterns for that keyword. This collects sentences into common groups. For example, one pattern is: `* i was *`, which can be read `any word phrase – I was – any word phrase`.

5. ELIZA picks a **reassembly** pattern to form a reply. If the program has several options for responses, it picks one at random. In our `* I was *` pattern, one response is "Perhaps I already know you were (2)." The number *(2)* in parentheses tells the program to substitute the word phrase that comes after `I was` in the sentence. If you typed in, "Then I was left at a bus station," the reply in this pattern could be, "Perhaps I already know you were left at a bus station." You might also get a more pragmatic, "Oh, really." It is important to know that ELIZA has no idea about the contents of phrases – it is just manipulating words to create sentences based on patterns.

6. ELIZA performs **postprocessing** substitutions of words. For example, it replaces the word *I* with *you*. If you type "I went to sleep," the program replies with, "You say you went to sleep?", which is the final reply rule after all the others are exhausted.

The data that controls ELIZA's personality is called a **script**. This gives the program all of the rules for forming replies. The script for the Rogerian therapist is called the **DOCTOR** script. It contains some greeting words, some final words when the program exits, a list of pre-substitution rules, a list of postprocessing substitution words, a list of synonyms, and a list of keywords with decomposition and reassembly rules.

Here is a sample rule for the `I am` keywords:

```
decomp: * i am *
reasmb: Is it because you are (2) that you came to me ?
reasmb: How long have you been (2) ?
reasmb: Do you believe it is normal to be (2) ?
reasmb: Do you enjoy being (2) ?
```

The program selects a random phrase out of the four provided. If I said, "I am afraid of cats," then this rule would be triggered, and one of these four phrases would be generated. It might say, "Do you enjoy being afraid of cats?" or "How long have you been afraid of cats?"

Almost all of the dialog created by ELIZA comes from the script file, making ELIZA a form of rule-based expert system, and also gives an open framework for creating new applications of conversational engines.

As all computer programs do, ELIZA's concepts for generating conversation were extended and amplified into a general purpose (that is, not just a therapist) software program called **ALICE**, which is a retronym for **Artificial Linguist Internet Computer Entity**. The simple script data file used by ELIZA became the **Artificial Intelligence Markup Language (AIML)**. ALICE was extended to be far more useful than ELIZA ever was and to cover far more subjects and capabilities. However, ALICE retains the same concepts of pre- and post-substitution, keywords, and response patterns. There is a Mycroft plugin engine that allows Mycroft to use AIML files to design conversations (`https://github.com/forslund/fallback-aiml`). This tool makes a great alternative construction kit for your robot personality.

Here, we've included an example of what AIML looks like (`http://www.digitalanywhere.com/projects/dany/index.html`). You will see blocks with categories of statements, and then a pattern of what that category looks like. The next tag, `<template>`, describes the response that the robot will make when this category is triggered. For example, the first category is about playing musical instruments. If you say to the robot "I play guitar," then the response will be, "Is guitar a game or a musical instrument?" We can create our simple sentence patterns this way:

```
<category>
<pattern>I PLAY *</pattern>
<template>Is <set_it><person/></set_it> a game or a musical
instrument?</template>
</category>
<category>
<pattern>I PLAY THE *</pattern>
<template>I wish I played an instrument. My parents could not afford
the lessons.
<think><set_it><set_does><settopic>play the
<person/></settopic></set_does></set_it></think></template>
</category>
<category>
<pattern>I PRIDE MYSELF *</pattern>
<template>It seems you have good self esteem.</template>
</category>
<category>
<pattern>I QUIT MY JOB *</pattern>
<template>Where did you work and why did you quit?</template>
</category>
<category>
<pattern>I READ *</pattern>
<template>I don't think I've read that. Can you give me a reference?
<think><set_it><set_does><settopic>read
<person/></settopic></set_does></set_it></think></template>
</category>
<category>
<pattern>I READ * BOOKS</pattern>
<template>Name some books in that genre.
<think><set_it><set_does><settopic>read <person/> books</settopic></
set_does></set_it></think>
<think><set_personality>critical</set_personality></think></template>
</category>
```

ALICE is **open source software (OSS)** released under the GNU public license.

Now, let's look at how we can add some context to our robot's conversations.

Adding context

One of the most important shortcomings of ELIZA is a total lack of memory. ELIZA only remembers the previous statement. You can't talk about your father and then say, "He was a good man." ELIZA will have no idea whom you are talking about. ELIZA has no concept of context beyond the previous sentence.

What is **context**? In the course of a conversation, we often shorten nouns into pronouns. We might say, "I like my dog," and in the next sentence say, "She is well behaved." Who does *she* refer to? We know it is the dog, but how does the computer know? We are going to add some ability to reason from context to our program.

We are going to create a storage object we will call the **context memory**. In that object, we will be able to keep up with several parts of our conversation, including the person we are talking to currently, the last subject we talked about, if we had asked any questions we still don't have the answer to, and the answers to any previous questions, in case we need them again. The computer will assume that a pronoun other than *I* will refer to the last subject, whatever that was. If I was talking about a dog, and then said, "She is so cute," then the robot would assume I meant the dog.

Previously, we discussed playing a game to get information from a person. The robot will be collecting and remembering this information, even after it is turned off, so that the next time that person is talking to the robot, it remembers the information it learned the last time – just as you do with a new friend. If you want to continue to expand this AI chatbot concept, you can use this information to create additional conversations. For example, if the human says they like baseball, the robot could ask what their favorite team is and then look up on the internet when the next baseball game is scheduled.

That is the end of our list of parts we are going to use to build our robot personality. We can now dive in and use our personality construction kit. I'm going to use the bones of the ELIZA Python open source program from Wade Brainerd to build Albert the robot's personality.

In the interest of time and space, I'm only going to put the parts here that I added to the base program. The entire code will be in the GitHub repository. You can get the original program at `https://github.com/wadetb/eliza` if you want to follow along beyond what's in this book.

Under construction

Let's review all the parts we have to put together to make our robot personality:

- Simulation
- Monte Carlo (stochastic, or random-based) modeling
- Our robot emotion state machine
- Perception of human emotion state machine

- Our robot biography (list of internal facts about the robot)

- A conversation engine framework called ELIZA

- Context or the ability to remember and return facts and "fill in the blanks"

In this and the next few sections, I will be presenting the code that I added to Albert to facilitate his artificial personality. A lot of it will be contained in script files that provide rules and patterns for Albert's speech. There will also be code functions for his emotion engine, human emotion model, and game engine. Let's get started:

1. I needed to add some new functions to the script language used by ELIZA. First, I added the context of our robot, who takes the place of ELIZA, the therapist. First, we have the opening words when we initiate the interactive conversation mode of Albert by saying "Hey, Albert" (or whatever you decided to call it), the wake word for Mycroft, and then just "Hello."

2. The robot responds with the `initial` phrase, as noted by the tag before the colon. We also have our closing phrase here. You can actually put as many phrases as you like, and the computer will randomly choose one. These rules go into the file I named `AlbertPersonality. txt`, which started as a copy of the original `doctor.txt` script file that came with ELIZA:

    ```
    initial: Hello. My name is Albert the Robot.
    initial: Hello. I am Albert the Robot, but you can call me
    Albert.
    initial: Hello. Nice to meet you. Call me Albert.
    final: Goodbye. Thank you for talking to me.
    final: Goodbye. It was nice to talk to you.
    final: Goodbye. I need to get back to my tasks.
    quit: bye
    quit: goodbye
    ```

3. I added some word substitutes in case the user calls the robot by name rather than *you*. This just substitutes *you* for anything you might call the robot. I also set synonyms for various versions of the robot's name, so you can call it *robot* or *bot*, as well as *Albert*, *Bert*, or even *Bertie*.

    A rule with `pre :` in front of it is substituted before any other processing takes place. From the first rule, if the word *robot* appears in a sentence, as in "Robot, how old are you?", the program removes *robot* and substitutes *you* to make the parsing consistent. We also change all uppercase letters to lowercase, so there are no capital letters in the rules. The `synon:` rule (synonym) replaces any of the listed words with the first word given:

    ```
    pre: robot you
    pre: albert you
    ...
    synon: you robot albert bert bertie bot
    synon: belief feel think believe wish
    ```

4. The next thing we needed was to create questions that we wanted the robot to ask in order to gain information. The program will automatically *harvest* this data in any of the keywords we define to appear in a sentence. Here is the definition for rules about asking questions:

```
questions:
reasmb: What is your name? <assert name>
reasmb: What can I call you? <assert name>
reasmb: How old are you? <assert old>
reasmb: How are you feeling today <assert feeling>
```

5. We create a new flag for questions to add to our script file. Each line represents one question, but we can ask it in different ways or forms. The program will select one version at random and decide which question to ask based on the relative priority we will set on the questions. The `assert` keyword with the `<>` symbols around it is another new flag I added to cue the context memory that we have created a context of asking some question, and the next statement is probably an answer:

```
datum: name
decomp * my name is * decomp I am *
decomp call me *
decomp <name> * # we are in the name context reasmb: Hello (1).
Nice to meet you
reasmb: Hi (1).
reasmb: Your name is (1), right?
reasmb: Thank you for telling me your name, (1) store: <name>
(1)
decomp * my name is *
reasmb: Hello (2). Nice to meet you.[welcome] [happy]
store:<name> (2)
```

6. I created a new data structure I called `datum`, the singular of *data*. This represents some information we want the robot to ask about. We give the datum a title – `name` in this case, as we want the robot to ask the name of who it is talking to. The `decomp` (for decomposition) tags are the patterns of sentences where the user might say their name. `*` represents any phrase. So, if the human says "Hello. My name is Fred Rodgers," then the robot will call them Fred Rodgers from then on. If the human says, "Call me Ishmael" for some unknown reason, then the robot will use that. We must reassemble the response phrase with the `reasmb` rules. `(1)` refers to the first `*` phrase that appears. If the user says, "I am John," then when we use the reassemble rules, `(1)` will be replaced by `John`. The robot will pick one of the phrases provided at random, such as: "Your name is John, right?"

I added another new tag to allow the robot to use the context memory area to perform two functions. We can declare a context subject when we ask a question, which we will do in a later section. For example, when we ask the user, "What is your name?", then we want the robot to know that the next answer will be in the context of that question. It's perfectly reasonable for the robot to say "Who are you?" and the user to immediately answer "Julia" with no other words in that sentence. How is the program to know what to do with *Julia*? The answer is the context flag, noted by being bracketed by < >. We read the decomp <name> * decomposition rule as if you are in the context of asking for a name, and you get a reply with no keywords, so take whatever you get as the answer.

Emotion tags are noted by a bracket, such as [happy] or [sad]. This will move either the robot's emotions or the robot's perception of human emotions, depending on whether it is a statement received from a human or a sentence uttered by the robot. There can be more than one emotion tag associated with a statement.

Here are the rules for listening for the answer to the age question:

```
datum: age
decomp <age> * I am * years old decomp <age> * I am % # integer
reasmb: You are (2) years old?
reasmb: (2) years old!
decomp <age> *
reasmb: You are (1) years old?
reasmb: (1) years old!
store: <age> (1)
```

On the final line, store: is the command to tell the computer that this is the answer to the question and to store that away in the dictionary with the title provided.

7. Next, let's use an example of an interaction with some emotions tied to it so that we can see how we will use the emotion engine to control what the robot says. This set of rules is enacted when the user says "How are you?" to the robot:

```
key: feeling
decomp: how are you feeling decomp: how are you
decomp: hows it hanging
decomp: how are you today
reasmb: <happy> I'm doing well. How are you? <assert feeling>
reasmb: <sad> I am feeling sad. How are you? <assert feeling>
reasmb: <curious> I am curious about my surroundings
reasmb: <friend> I am feeling friendly today
reasmb: <welcome> I am in a welcoming mood today, my friend
reasmb: <frust> I am a bit frustrated, to tell you the truth
reasmb: <frust> I am feeling a bit frustrated
reasmb: <strange> I am having relationship problems
reasmb: <distant> None of my friends have come to visit
reasmb: <tired>  My batteries are low.  Maybe I need a rest.
```

8. We will be putting the robot's emotions into the context memory so that the script processing program can have access to it. We treat an emotion – for dialog purposes – as part of the context that we are speaking about, which I think is a reasonable approach to working with emotions. Each emotion has a tag or name in the context memory dictionary. If the predominant emotion in the robot is happy, then the robot sets the `happy` context in the context memory. Then, the rule base will use the context tag to determine which phrase to use to reply to "How are you feeling?" We can also ask a follow-up question. Look at the rule for `<happy>`. The robot replies, "I'm doing well. How are you?" and then sets the `feeling` context to let the engine know that we asked a question about feeling. Finally, the last line relates to the `tired` emotion. If the robot is feeling tired, then we jump to a separate section for the robot to talk about being tired. We make it a separate routine because we need to call it from several places, which illustrates the utility of this rule-based approach to speech. I don't want to imagine how many C or C++ source lines of code it would take to create all of these rules for each line of dialog. We continue to modify the scripts using these guidelines until we have completed all our questions and have patterns for all of the answers.

Now, we are going to switch back to Python code for the rest of this example. Our next section describes how we are going to simulate emotions.

Developing the robot emotion engine

Now we are going to put together the robot's emotion model. This is the heart of the artificial personality as it computes, updates, and remembers the emotional state of the robot. The robot starts in a generally neutral state and performs updates based on the combination of eight emotional traits: *happy/sad, welcoming/distant, friendly/ stranger, curious/frustrated*, and *fresh/tired*. As events happen in the experience of the robot, it gets cues that cause its emotional state to change. For example, if the user said "That is stupid" to something the robot said, then the robot would add to the `sad` axis of its emotion.

We compute the overall emotional state using polar coordinates, just like you saw in the diagram we drew earlier in this chapter (*Figure 9.4*). The current emotional state is determined by computing the center of mass of the other emotions. If the emotions are more or less balanced, the center of mass of the emotions, as plotted on our polar chart, would be near the center. If the robot is mostly happy and friendly, then the mass moves more over to that side of the graph. We pick the single emotional state that is closest to the center of the mass. This is intended to be the basis of creating a complex emotional character for the robot. The attribute of *fresh/tired* is unique, in that the value for that emotion is based on the elapsed runtime of the robot.

The primary expressions of emotion for the robot will be the position of the robot arm – happier robots carry their arms higher and more forward – and the choice of vocabulary in conversation.

The following block of code creates data entries for the emotion engine in the format required by our conversation engine. We are creating a Python interface to the rule-based format so that we can connect it to the rest of the robot:

```
class RobotEmotionEngine():
  def __in_(self):
    self.emostate = [90,0]
    self.emoText = "neutral 50"
    self.emotions = {"happy" : 50, "sad": 50,"welcome" :
50, "distant":50,"friend" : 50,"strange" :50, "curious" :
50,"frustrated":50, "fresh" : 50, "tired",50}
    self.bio = {"name":"Albert Albert", "lastname":
"Albert", "age": "6 months","maker": "granddad",
"color":"green","food","electricity","author":"Isaac Asimov,
of course","school": "I do not go to school but I love to
learn","hobby":"picking up toys", "job":"picking up toys"}
    # list of happy emotions and sad emotions self.
emotBalance={"happy": "sad", "welcome":"distant","friend":
"strange", "curious": "frustrated","fresh": "tired"} self.
emotionAxis{"happy":112, "welcome": 22,"friend":67,"curious":157,
    "sad":292,"distant":202,"strange":247,"frustrated",337}
    self.update()
  def change(self,emot, val):
    self.emotions[emot]=val
    balance = 100 - val
    otherEmotion = self.emotBalance[emot]
    self.emotions[otherEmotion]=balance
```

Next up is the update function; this function checks to see if we've had a change in our emotional state, and if so, we change our current emotion:

```
def update(self):
    rmin = 100
    rmax = 0
    thetamin =360
    thetamax=0
    for emote in self.emotions:
        theta = self.emotionAxis[emote]
        thetamax = min(theta,thetamax)
        thetamin = max(theta,thetamin)
        r = self.emotions[emote]
        rmin = max(rmin, r)
        rmax = max(rmax,r)
    stateR = (rmax-rmin)/ 2
    stateTheta = (thetamax-thetamin) / 2
    for emo in self.emotionAxis:
```

```
        thisAngle = self.emotionAxis[emo]
        if stateTheta > thisAngle
        myEmotion = emo
        break

    self.emostate = [stateTheta, stateR]
    if stateR < 55 and stateR > 45:
        myEmotion = "neutral"
    self.emoText = myEmotion + " "+ str(stateR)
    print "Current Emotional State"  = myEmotion, stateR,
stateTheta
    return
```

The robot also needs a model of the human it is talking to so that it can make different responses based on how the human is interacting. In the next section, we are going to create a smaller version of the emotion model we used earlier.

Creating a human emotion model

We model four emotions for our human interactions for the robot to use in formulating responses: *happy/sad* and *welcoming/distant*. We can put emotion tags into our patterns in the script file with [happy], [sad], [welcome], or [distant] to mark the emotions of responses. For example, if we are not getting answers to our questions, we can mark that response with [distant] to note that our subject is not being cooperative:

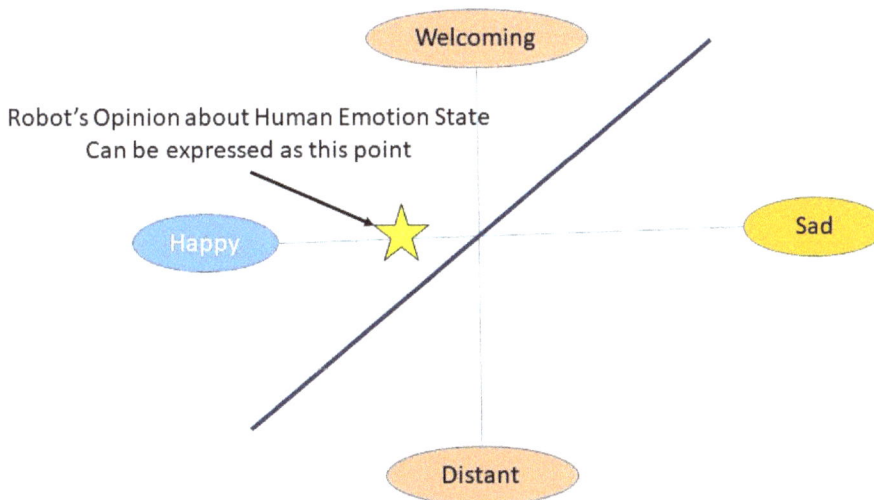

Figure 9.6 – Simplified human emotional model showing the axes
between happy and sad emotions: robot point of view

Our human emotion model makes use of a Python dictionary data structure to hold our model. Let's take a closer look at this:

1. We have two axes: the *happy/sad* axis and the *welcoming/distant* axis. We move the *happy/sad* index up or down based on responses. If we think a response expresses happy thoughts ("Do you like school?" "Yes"), the program moves up the emotion index in the *happy* direction. We use the intersection of these to set the current emotional index. If the human is near the center, we note this as neutral, our starting point:

    ```python
    class HumanEmotionEngine():
      def _init_ (self):
        self.emostate = [90,0]
        self.emoText = "neutral 50"
        self.emotions = {"happy" : 50, "sad": 50,"welcome" : 50,
    "distant":50}
        # list of happy emotions and sad emotions
        self.emotBalance={"happy": "sad", "welcome":"distant"}
        self.emotionAxis = {'distant': 315, 'welcome': 135, 'sad':
    225,'happy': 45}
        self.update()
    ```

2. Let's look at the change function. If happy goes up, sad goes down, so we balance this automatically when emotions change:

    ```python
    def change(self,emot, val):
        self.emotions[emot]=val balance = 100 - val
        otherEmotion = self.emotBalance[emot]
        self.emotions[otherEmotion]=balance
    ```

3. The update function gets the current emotional balance of the human model:

    ```python
    def update(self):
        stateR = self.emotion["happy"]
        stateS = self.emotion["welcome"]
        self.emostate = [stateR, stateS]
    ```

4. If the emotional state is near the middle, we call that neutral:

    ```python
    if stateR < 55 and stateR > 45 and stateS < 55 and stateS > 45:
     myEmotion = "neutral"
    happySad = stateR-50 welcomDist = stateS-50
    if abs(happySad) > abs(welcomDist):
     myEmotion = "sad"
    if happySad > 0:
     myEmotion = "happy"
    else:
    ```

```
      myEmotion = "distant" if welcomDist> 0:
      myEmotion = "welcome"
      self.emoText = myEmotion + " "+ str(stateR)
      print "Current Human Emotional State" = myEmotion, stateR,
      stateTheta
      return
```

This next section discusses where we are going to store the information we are collecting from humans.

Creating human information storage

This data structure stores the information we have collected about the human user and lets the robot know what questions have yet to be answered. We use a Python dictionary to hold the data. Let's make a freeform version of a database-like table. I add values to the dictionary as we go to make extension easier. I put a list of miscellaneous questions called `stuff` to throw some spice into the robot's information gathering so that the questions don't seem too monotonous. The robot will ask if you like the color pink, singing, dancing, robots, airplanes, and so on.

We put the list in priority order by copying the dictionary and replacing the data fields with relative point values from 20 to 0. name is first with 20 points, age is second with 18, and so on. As each question is answered, we set this point value to zero. For example, if we get an answer for "What is your favorite food?" as "apple," we set `self.info["food"] = apple`, and set `self.points["food"] = 0`. I also added some extra questions for things that my grandchildren like, with the points set very low so that the conversation is more varied. I ask about princesses, airplanes, dinosaurs, singing, and building things:

```python
class HumanInformation():
  def __init__(self):
    self.info = {"name":"none"}
    self.info["age"]=0
    self.info["school"]="none"
    self.info["feeling"]="none"
    self.info["food"]="none"
    self.info["book"]="none"
    self.info["subject"]="none"
    self.info["song"]="none"
    self.info["teeth"]="none"
    self.info["jokes"]="none"
    # stuff is random information that we use to get more information
and have the human answer questions
    # these are aimed at 3-7 year olds
    self.info["stuff"]="none"
    self.stuff = ["the color pink", "singing", "dancing", "dinosaurs",
"race cars", "building things",
```

```
    "robots", "airplanes", "space ships", "unicorns", "princesses"]
self.points = self.info
    # setup points scoring scheme
    points = 20
    for item in self.points:
      self.points[item]=points
      points -= 2
```

Now, in the next section, we'll create a memory for our robot so that the robot can remember what it has already learned. We need a place to put the answers we receive.

Context memory

This block of code constructs the robot's context memory. You can think of this as a pool of information that the robot can remember. We set up the robot's emotion and the model of human emotion to both be neutral. I created some data structure so that we can refer to multiple human users by pointing `self.currentHuman` to the `HumanInformation` data object we created previously.

This is where we have the robot remember what emotional state it is in, along with what the software perceives as the human's emotional state so that the robot doesn't suddenly forget the decisions it has already made. In the following code snippet, we define `self.emotion`, which is the robot's internal state, and `humanEmotion` for the person we are interacting with. Then, we use these structures to create a file we write to disk so the robot's personality persists if you turn it off. We use the `inContext` function to retrieve data from the context pool. If no data is available, we return an integer of 0:

```
class ContextMemory():
    def __init__(self):
        self.currentContext = "None"
        self.currentHuman = None # pointer to the data file for the
human we are currentl talking to
        self.humanFile = []
        self.emotion = "happy"
        self.humanEmotion = "happy"
        self.contextDict={}
        self.contextDict['currentHuman'] = self.currentHuman
        self.contextDict['robotEmotion'] = self.emotion
        self.contextDict['humanEmotion'] = self.humanEmotion

    def inContext(self, datum):
        if datum in self.contextDict:
            return self.contextDict[datum]
        else:
            return 0
```

```
    def setHuman(self,human):
        self.currentHuman = human
        self.humanFile.append(human)  # add this person to the
database of people we know

    def addHuman(self,human):
        self.humanFile.append(human)
```

In these sections, we created a place for the robot to store information about how it currently feels, what emotional state the robot is in, and what it has learned about the human it is talking to. We also have a structure to add new humans to our database for when we meet someone new.

Summary

This was a very busy chapter. We created an emotional engine for our robot and created a conversational interface so that it can interact with people. We introduced simulation concepts, as we were creating a simulation of emotion for our robot. We can't create real, biologically motivated emotions, so we simulated emotions via playing a game. In this game, we tried to get humans to provide information about themselves, which is what we humans call *small talk*. We assigned point values for emotions and created an emotional state machine both for the robot's internal *feelings* and a representation of how the robot perceives your emotions via interactions.

In our final chapter, we will talk about your robotics educational journey, give some guidance on careers in robotics, and talk about the future, which is always a perilous topic.

Questions

1. What is your favorite movie robot? How would you describe its personality?

2. What techniques did the movie-makers use to express that robot's personality (body language, sounds, and so on)?

3. What are the two types of chatbots? List some of the strengths and weaknesses of each.

4. In the diagram of modeling custom distributions (the airport example), the bottom image shows two standard distributions and two uniform distributions. Why don't the curves go all the way to the top of the graph?

5. Design your own robot emotions. Pick six contrasting emotions that can express the entire range of your robot's personality. Why did you pick those?

Further reading

- *Affective Computing* by Rosalind Picard: `https://direct.mit.edu/books/book/4296/Affective-Computing`. This book is a foundational text in the field of affective computing, which is directly relevant to simulating emotions in AI.

- *Emotion Modelling for Social Robots* by Ana Paiva, Iolanda Leite, and Tiago Ribeiro: `https://people.ict.usc.edu/~gratch/CSCI534/Readings/ACII-Handbook-Robots.pdf`. This book includes research papers and studies focused on implementing emotional responses in robots.

- *Designing Sociable Robots* by Cynthia Breazeal: `https://direct.mit.edu/books/book/2309/Designing-Sociable-Robots`. A key resource for understanding the integration of emotional and social dynamics in robots.

10

Conclusions and Reflections

We have been on quite a journey throughout this book. I've learned a lot, and I hope you have as well. I've had the chance to revisit my love of robotics and spend a lot of time examining the state of the art of **artificial intelligence** (**AI**) and robot design to try and find a way to explain the concepts to you in an easily digestible form.

In this chapter, we will discuss finishing our robot. I'll provide you with some advice on careers in AI and robotics. We will also talk a little about the future of AI, at least as I see it, and finish with a discussion about risk.

The following main topics will be covered in this chapter:

- Learning when to stop
- Careers in robotics
- Exploring the current state of AI
- Understanding risk in AI

Learning when to stop

Over the last nine chapters, we've worked on designing a specific robot to perform a specific task. We designed *Albert* the robot to pick up toys in an unstructured environment, namely a family home. To do this, it needed to be able to recognize toys with a camera, pick up toys with a robot arm and hand, navigate inside a house, and deposit toys in a toy box. We also added interaction, teaching the robot to listen and react to commands. Finally, it received an artificial personality and simulated emotions.

So, the next question is, are we finished with designing, building, and testing our robot? Sometimes, the most difficult part of designing, building, and testing a robot is determining when it is finished. Quite often, I see that some little thing might be improved, or some detail added, or some feature enhanced. Oh, the robot needs a spotlight. It would be nice if it could remember what toys are in the toy box. What if it had two arms? And so on. You can continue in this vein forever, tinkering and adding, never quite finishing anything.

The solution to this is to set specific goals and measure against them. For example, we want the robot to pick up toys. How many toys? All of the ones that are on the floor. So, we can run several trials and, if at the end of each test, all of the toys have been picked up, that is finished. But what if not all of the toys have been picked up? What is good enough? How about we say if we can pick up any remaining toys in one pass by holding them all in our hands, then that is acceptable. What level would not be acceptable? Well, picking up no toys at all is not acceptable. What about half? Would we consider that OK? I'd say probably not. You can continue this self-conversation until you say with some clarity where the finish line is. Then, once you've crossed it, you're done.

Sometimes, we have to temper expectations when we're trying something new or innovative. Expect to spend some time ironing the bugs out. I've had the experience of creating something and having the robot do something new and unexpected that ended up being far better than what I started with. For example, I created a robot that had a fairly sophisticated *follow me* function. It used body recognition to identify humans, and then, when commanded, followed behind one person, even through a crowd – so long as it could always keep the person it was following in its field of view. The robot was programmed to keep a six-foot distance from the person. This means if you walked toward it, instead of away, it backed up. You could then steer the robot backward in any direction simply by walking toward it – my *follow me* function became a *walk ahead* function.

Other times, an innovation just does not work, and at some point, it must be abandoned. Generally, you can tell this when you have to keep adding *crutches* or workarounds over and over again. The software becomes more complex and fragile with each workaround. I had this problem with a sonar-based obstacle avoidance system – the sonar sensor I used was just too unreliable and was very sensitive to the surface involved – for example, it could not see (get any echoes) from polished wooden doors. After a few weeks of testing, we abandoned that sensor and went with another light-based sensor called an **Infrared Proximity Detector** (**IRPD**) that worked much more reliably.

In the next section, we'll take a look at some career paths that you could take if you are interested in robotics and/or AI.

Careers in robotics

I am often asked what sorts of skills or degrees robot designers need to have, or what courses they should take. I meet a lot of young people at robot competitions, student events, conferences, and recruiting trips. A lot of the advice I give people I have put into this book already – especially now that AI, neural networks, **graphics processing units** (**GPUs**), expert systems, chatbots, navigation, and image processing are all important. You need to understand that *robotics* is an umbrella term that covers a lot of industries and a lot of skill sets.

Who uses robotics? The range of applications is continuing to grow every day. These include the following:

- In the medical field, there is robot-assisted surgery, robotic prosthetic limbs, exoskeletons helping paraplegics to walk, and implantable devices to help people hear and see.

- We have self-driving cars and self-flying airplanes and helicopters in transportation (which is what I do).

- We have robots delivering packages on the sidewalks in San Francisco, and a few companies testing parcel delivery by aerial drone in Africa, Switzerland, Texas, and other places.

- We have self-driving cars being tested in several countries. Safety features that debuted in the *DARPA Grand Challenge* robot car race in the United States and were developed for autonomous cars – lane keeping, adaptive cruise control, driver assistance, and self-parking – are now common features on even base-level automobiles. There are over 80 companies currently developing some sort of electric **vertical takeoff and landing** (**VTOL**) manned vehicle, every one of which uses advanced automation and autonomy as part of its control system. In far western Australia, Rio Tinto Mining has developed the *Mine of the Future* in Pilbara, where 80 autonomous trucks are remotely operated from Perth, 1,500 kilometers away.

The future of robotics is just being written, and you, reading this book, may play a part in determining the path it takes.

So, what skills are required to design and make robots like the ones I just described? The truth is that a modern robot company would employ just about every skill imaginable. Let's look more closely at some of these skills:

- Even the simplest robot takes **mechanical designers** to develop the parts, gears, and levers to make robots move and help package the electronics.

- **Electrical engineers** work with batteries and motors.

- **Radiofrequency** (**RF**) **engineers** and **technicians** work with radios and datalinks that are used to connect mobile robots to their control stations (such as an **unmanned aerial vehicle** (**UAV**)).

- **Cognitive specialists** design AI routines, develop robot emotions, and harness machine learning techniques, as well as design user interfaces.

- **Writers** and **artists** craft voice routines, write dialogue, design user interfaces, write manuals and documentation, and add creative touches to the inside and outside of the robot.

- **Managers** and **supervisors** track budgets and schedules.

- **Supply specialists** work with suppliers, parts, build-to-print shops, electronics warehouses, and salesmen to get the parts to put together the assembly line.

- Industrial robots are managed by special types of **programmers** who use **programmable logic arrays** (**PLAs**) and **ladder logic** to control robot arms that paint and assemble components. This type of programming emulates how robots were originally designed using relays and switches.

- **Bookkeepers** and **accountants** make sure the bills, as well as the employees, are paid.

- **Salespeople**, **marketing**, and **customer relations** teams get the product sold and keep the customers happy.

All of these skills have to be present in some form in a professional roboticist, particularly if you think you will run your own company. I've been part of every size of robot project, from one person to thousands. Each has its strengths and weaknesses, but as a robot designer, you can be sure that you will be the center of the storm, making the magic happen, solving problems, and turning ideas into physical form. To me, there is no more satisfying moment than seeing my creation driving or flying around, doing its job, and knowing that all the hard work, research, coding, mechanics, sleepless nights, smashed fingers and toes, and skipped meals were worth this result.

Now, let's talk a bit about the hype surrounding robotics and AI that you have probably seen in the news.

Exploring the current state of AI

There is a lot of hype going on right now in the intersecting worlds of AI and robotics. And a lot of it is just exaggeration.

One common myth is that robots are taking jobs away from people. In truth, robots and automation free up workers to do more productive tasks. The truth of this can be seen in job statistics – unemployment in the US is at a 50-year low (`https://www.wsj.com/articles/january-jobs-report-unemployment-rate-economy-growth-2023-11675374490`), despite massive improvements in factory automation. However, according to *The Harvard Business Review*, the improved productivity of robotics creates more jobs than it removes (`https://hbr.org/2021/03/why-robots-wont-steal-your-job`). The overall level of employment has *increased*, not gone down because of automation and increased productivity.

I do recognize that the modern worker, even someone like myself, who works in technology, must be ready and willing – at any age – to retrain themselves and to learn and adapt to new ways of working, new economies, and new opportunities. I've had to completely retrain myself at least six times as new markets were invented and new technologies emerged. Sometimes, there is a *second wave* where some technology was invented but then disappeared when it was too expensive for the benefits it provided, or the proper hardware had not been invented yet. Neural networks fit into that category, as does virtual reality, which was a big deal in 1999, and has now re-emerged with very small high-resolution screens that were developed for cell phones.

Looking ahead in AI and robotics

I'm quite interested in the long-term impact of what has been called the *sharing economy*, where companies such as Uber, Lyft, and Airbnb create value by connecting suppliers and consumers on a massive scale without owning any of the capital or resources to provide any services. All of this is enabled and made possible by the ubiquitous internet, which continues to grow and evolve at a rapid pace. The availability of the internet has allowed the general public or an individual student to have access to supercomputer-level capabilities to run AI programs such as ChatGPT online, which are too large to fit in a home computer, smartphone, or tablet. I often use the term, "*but that's a decade in internet years*" while referring to some idea that is maybe 24 months old to indicate the rapid turnover in internet tech. This trend will continue. It will be interesting to see if anyone owns a car in 20 years, or only a subscription to a car service.

Another trend that has become very interesting is the lowering of barriers to entry in a lot of businesses. You used to have to have an enormous machine shop and giant machines to make precision machine parts – before 3D printers came and put that capability on your desktop. **Generative AI** is AI that can synthesize writing and drawing and can also directly create music, write programs and software, provide advice, and assist users in writing scripts, drawing pictures, and making animations powered by only a text prompt. Do you want to make movies? You can do so on an iPhone. Do you want to start a recording studio? The parts for professional results (with a large amount of effort) are available for less than $200 and you can use AI to generate lyrics, chord progressions, arpeggios, or even song ideas.

One item that fits into that category of lowering barriers to entry is drones or small UAVs. When I started making UAVs, a decent **global positioning system** (**GPS**) and **inertial measurement unit** (**IMU**) – the things that make unstable quadcopters possible to control – cost tens of thousands to hundreds of thousands of dollars. The real breakthrough in drone technology did not come from aviation, but rather from our cell phones. The developments in cell phones enabled companies to invest billions of dollars in making the next cell phone, smartphone, hand-held computer pacifier, or whatever you would want to call it. The convergence of very small radios, very small GPSs, and very, very small accelerometers enabled an entire world of unmanned flying objects – quadcopters, gliders, airships, airplanes, and hybrid VTOL craft – to emerge. That, along with higher density batteries that came from (you guessed it) cell phones and laptops, allowed people to discover that if you put enough power on it, you can make almost anything fly, including you.

The secret to the flying quadcopter's amazing success is that the tiny accelerometers (which measure changes in movement) and tiny gyroscopes (which measure orientation changes) became cheap and readily available. Without these sensors, and the robotics algorithms that control them, quadcopters are unstable and impossible to control. Another reason for the quadcopter's success is that it uses only the throttle setting – the speed of the motors – to control all its aspects of flight, including stability. This compares with the very complicated collective controls and cyclic pitch controls that make a helicopter work. You can see the difference between a radio-controlled helicopter, which is very expensive and only a few people can fly, and a quadcopter, which is quite cheap and can be flown by anyone, with the help of a computer and some sensors. You can add a drone autopilot to a collective/cyclic radio-controlled helicopter and end up with a very controllable drone helicopter. Quadcopters and more complex flying machines use AI for stabilization, adaptive flight control, object recognition, and obstacle avoidance.

The other side of these advances in AI and robotics has also resulted in a backlash from some people and can be described as robot phobia.

Is AI phobia reasonable?

You have probably seen some blazing headlines on the internet from various very credible sources saying some incredible things.

About a decade ago, Stephen Hawking, a leading scientist, stated that "*The development of full artificial intelligence could spell the end of the human race... It will take off on its own and re-design itself at an ever-increasing rate... Humans, who are limited by slow biological evolution, can't compete, and will be superseded*" (`https://www.bbc.com/news/technology-30290540`). This quote is frequently used even today by critics of AI.

More recently, Elon Musk suggested that AI could lead to *civilization destruction*, although he has invested significantly in the growth of AI (`https://edition.cnn.com/2023/04/17/tech/elon-musk-ai-warning-tucker-carlson/index.html`).

Bill Gates, former chairman of Microsoft, takes a more middle ground, stating that AI presents both promise and concerns. In an open letter, he elaborated on AI's potential but also discussed the risks of developing this tech. He wrote, "*The world needs to establish the rules of the road so that any downsides of artificial intelligence are far outweighed by its benefits*" (`https://www.forbes.com/sites/qai/2023/03/24/bill-gatess-open-letter-suggests-ais-potential-is-both-exciting-and-terrifying/`).

First, let me clarify that I do not think I'm in the same league as these gentlemen. But what I am is someone who works every day at the leading edge of AI and unmanned vehicles and robotics, and who attends and speaks at a lot of AI conferences. The source of this fear of AI can be found in any bathroom, hanging on the wall – we are using our mirrors to reflect our motivations and desires onto our creations.

I would liken the state of the art of AI today to a teenager who has managed to strap a V-8 engine to a skateboard – an enormous amount of power with no way to use it. While computers, hardware, and networks have advanced and scaled at an amazing rate, we are still building software the same way today as we did 20 years ago, and many of the advances of AI were invented 40 or 50 years ago. For instance, neural networks were invented in the 1940s (`https://link.springer.com/article/10.1007/BF02478259`). Genetic algorithms were suggested by Alan Turing in the 1950s. We have been able to make some advancements in applying learning theory now that we have lots of data (thank you, internet) and lots of bandwidth, but we have not created any startling new concepts or capabilities that show any sort of breakthrough in AI. Recent work in generative AI –such as **ChatGPT** and **Dall-E** image generation – has created what might be described as a computer version of an imagination (creating data stepwise out of noise), which indeed is a step forward, but still, a generative AI such as ChatGPT is doing what neural networks did before: predict the next word in a sentence.

What is generative AI? In this case, we can ask the AI itself for a definition. The following text was written by an AI program called ChatGPT4 in response to the prompt, "*Can you provide me with a definition of generative AI?*":

> "*Generative AI refers to a subset of artificial intelligence technologies that can generate new content, such as text, images, music, or code, based on the patterns and information it has learned from its training data. This type of AI uses machine learning algorithms, particularly generative models such as generative adversarial networks (GANs) or transformers, to understand and replicate the complexities of human-generated content.*
>
> "*The "generative" aspect implies that the AI isn't just analyzing or processing existing information but actually creating something new and previously unseen. For instance, in the case of text, generative AI can write essays, poems, or code; for images, it can create artworks or realistic photos of things that don't exist; in music, it can compose new pieces in various styles.*"

But is this what is referred to as **general AI** (software with human intelligence)? I don't think so. While the hardware and software are advancing, I simply don't see any path forward that leads to the type of problems that the esteemed Mr Gates, Mr Musk, or Dr Hawking suggested. Why not? We'll focus on this in the next section.

Comparing the needs of humans and AI

The reason why I don't see significant problems is because robots don't have needs. Humans have needs and ambitions to exist. We are encased in a frail flesh cell, what William Burroughs called the *Soft Machine*. We must provide air, food, water, shelter, and clothing to protect our fragile shells and interact with other soft machines (people) to reproduce and make more of ourselves. You can argue, as Richard Dawkins did in his book *The Selfish Gene*, that all of this is simply an evolved way for our DNA to perpetuate itself, and that we are simply the product of our biological programming. It is impossible to separate a human from their needs – if we don't, we die in a matter of minutes. It is our needs that drive us forward, to come out of the trees, to learn to farm, to build cities, and to make civilizations.

Robots, on the other hand, do not have needs as a condition of their existence. They are just sets of instructions we have set down in electronics – golems with words in their heads that make them move (as described in the book *Feet of Clay*, by Terry Pratchett). If we don't provide food to them – nothing happens. If we don't use them – nothing happens. If we forget them for a week and check on them later, they are still the same.

First, let's discuss what humans' needs are. Maslow came up with the **hierarchy of needs** back in 1943, and he has been quoted ever since. Maslow says that we not just have needs, but they form a hierarchy – the more important needs are at the bottom while the more abstract needs are at the top. We only worry about the need at any given level when all of the needs below it are satisfied, as shown in the following diagram:

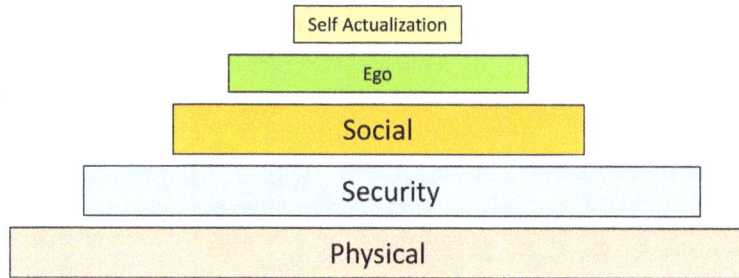

Figure 10.1 – Hierarchy of needs for humans

Let's look at this pyramid of needs in detail:

- At the bottom are the *physical needs* – air, food, water, and clothing.

- The next level is *security* – we need to feel secure from predators or from other humans wanting to harm us.

- Above the security needs are *social needs* – to be in a group or part of society. Humans want to belong to a family, a community, a nation. This drive is very strong, as our invention of war to protect our society attests.

- Next, we have *ego needs* – the need to be recognized, to be special, to stand out from the crowd we fought so hard to be part of. Remember, we only get to express this once all the other needs are taken care of, so you only worry about recognition once you are part of a group.

- Our final need is called *self-actualization*, as described by Maslow – we would call it self-improvement, or the drive to improve one's self. This is where we get athletes, artists, musicians, and people who write books.

> **Note**
>
> It has been a running joke between my wife and me that every textbook we ever read in college contained some reference to Maslow and his hierarchy of needs. This was quite unusual since I studied math and engineering, and my wife's degree is in human resources. I appreciate the irony that this book now adds to that list of books that reference Maslow.

Now, let's look at machine intelligence and imagine for a moment what a set of robot needs might look like. I found this an interesting thought experiment – we make a baby robot that is fully capable of learning anything a baby human (or baby mouse, or baby cricket) can. What needs would it have? Let's look at a modified version of Maslow's hierarchy of needs:

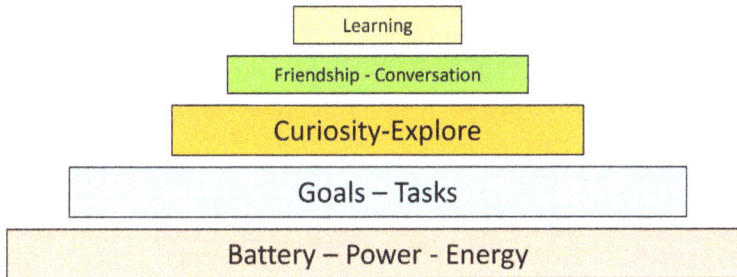

Figure 10.2 – Hierarchy of needs for robots

We can break this down as follows:

- In contrast to a robot's needs, hunger is built into humans' biology. However, in the case of an AI system, we, the creators, would need to build it in. That would equate, as we did in our artificial personality, to *electrical power* or *battery life*.

- The next level of needs would be the *goals* and *tasks* for which the AI was created.

- The next level up could potentially be *curiosity* and the need to explore – our AI system would have a drive to acquire more data, or to get access to more resources. Once a robot has data, this gives it the basis to get more data, and so on.

- The next level of needs we would endow to our AI would be the need for *friendship* or *communication*, either with other robots or with people.

- Finally, we could give our robot the need for *learning*, or to acquire and learn new skills and techniques – to grow as a robot.

You may have noticed that we did not cover these subjects in this book, nor any other. We did not talk about giving the robot needs, only how to simulate emotions and some rules for conversation that make no sense to the robot at all. It gets no joy from telling a terrible joke to a 7-year-old because it does not know what any of those concepts are. It just sends electrons down one path or another because we tell it to. The only intelligence here is the reflection and imagination of us, the robot designers. Everything else is an illusion – a magician's trick.

I get this sort of question quite a lot and felt that I could give you some of my answers if that helps arm you against the critics of AI. The bottom line is that I simply do not worry about AI taking over the world. I don't mean to say that there will never be a general AI, just I don't see one coming about in the foreseeable future.

With this context under our belt, let's discuss how to understand and manage risk in AI.

Understanding risk in AI

One subject I talk about frequently at conferences and in print is the risk of AI in terms of *trust* and *control*. I'm not talking about AI running amok here, but rather how to make AI dependable. It is quite interesting that the sort of AI we have been considering – specifically, ANNs – does something very few other computer software do. Given the same inputs and conditions, the output of an AI system is not always the same. Given the same inputs, the AI system will sometimes come up with a different answer. The formal name for this behavior is **non-determinism**.

There is a second corollary to this. Given the same inputs, the AI process will sometimes take a different amount of time to complete its task. This is simply not normal behavior for a computer.

Admittedly, we are not using AI to get answers to math questions such as *2+2*, but rather how to do things such as diagnose a cancer tumor or recognize a pedestrian in a crosswalk for a self-driving car. How do we deal with a computer output that may be wrong? You can verify this for yourself – look at the examples we covered when we performed training on neural networks. Did we ever achieve 100% success from a training run, where we got all of the answers right? No, not once. This is because ANNs are *universal approximation functions* that map inputs – which can be quite complex – to outputs. They do this by dealing with probabilities and averages, which were developed over time. You can think of an artificial neuron as a probability engine that says, *45 out of the last 50 times I got this set of inputs, the output was supposed to be true. The odds are it will be true this time*. And it sets itself to true. We may have millions of little artificial neurons in our network, each of them making the same sort of calculation. The net result is making a very educated guess about the answer.

For most applications of our neural networks, this is acceptable behavior. We are classifying pictures, and it is acceptable if a few are wrong. We do a Google search for platypus, and we get one picture out of 100 Platypus brand tennis shoes. That is OK for a Google search, but what if we were doing something more serious, such as recognizing pedestrians in a self-driving car? Is it OK if we misidentify one pedestrian out of 100 and don't avoid them? Of course not. That's why, right now, we don't allow AI systems in such critical functions. But people want to use AI in this way – in fact, quite a lot. It would be great to have an AI system that recognizes geese in flight and tells your airliner how to avoid them. It would be great to have an AI system recognize that a patient was misdiagnosed in the hospital and needed immediate attention. But we can't do that until we come up with processes for dealing with the non-deterministic and thus non-reliable nature of AI.

Currently, we deal with non-deterministic elements in automobiles all of the time. They are called drivers. It is widely believed that the vast majority of car crashes are caused by the human element, which is why we need self-driving cars with a better percentage. How do we deal with human drivers? Let's look at the necessary criteria for a driver:

- We require them to be a certain age, which means they have gained experience

- They have to pass a test, demonstrating competency in accomplishing tasks

- They have to demonstrate compliance with rules and regulations by passing a knowledge test

- They have to get periodically re-certified by renewing their license
- We also require seat belts and airbags to partially mitigate the risk of the human driver making mistakes by reducing some of the resulting injuries

We can apply these types of criteria to AI. We can require a certain amount of training cases. We can test and demonstrate a level of competency. We can predict the level of errors or mistakes in advance and put measures in place to mitigate that risk. Perhaps we can have two AI systems – one that detects obstacles and another that has been trained to recognize that the first AI has made a mistake. If we have a 90% chance of the first AI being right, and another 90% chance of the second AI being right, then we have a *90% + (90% of 10%) = 99%* chance of avoidance.

I think the key to using AI in safety-critical applications is being able to predict risk in advance, and designing in advance to mitigate either the cause of the risk or the effect.

Summary

In this final chapter of this book, we summarized our journey through robotics and AI. We talked about robotics as a career and discussed AI robotics as a profession. I brought up some issues regarding the future of AI, both real and imaginary. Drones and self-driving cars are real; robots taking jobs from humans or taking over the world is imaginary, at least in my opinion. I talked about robots and AI not having needs, and thus lacking the motivation, pressure, or even capability to evolve. Finally, we talked about risk in AI and how to recognize it. I hope that this information gives you some guidance in your interest in robotics and AI and provides some *insider information* from a practitioner in this area.

Now that we're almost at the end of this book, I want to thank you for coming on this journey with me. I hope you have learned something along the way, even if it is just to know more questions to ask. I encourage you to dive in and build your own robot, learn about AI, and become part of the community of people who contribute to robotics as a hobby or a profession.

I have to acknowledge a debt of gratitude to all of the robotics and AI open source communities for making all of this material, knowledge, and expertise available, and continuing to make AI the poster child for why open source, as a model for the advancement of human knowledge, works and works well. ROS, which is entirely run by volunteers, is a case in point as it makes building robots so much easier.

Questions

1. Given that we started the chapter on a light note and ended up talking about robot phobia and philosophical questions about existence, do you feel that AI is a threat? Why or why not?
2. List five professions that would be necessary to turn our Albert robot into a product company.
3. If we imagine a company that was going to put Albert the robot into production, would it need a psychologist? For the robot, or for the humans?

4. What components found in cell phones or smartphones are also found in quadcopters?

5. Why are AI systems, specifically ANNs, naturally non-deterministic in terms of results and time?

6. What might be a practical application of an AI system that predictably makes mistakes?

7. If an AI system was picking stocks for you and predicted a winning stock 43% of the time, and then you had a second AI that was 80% accurate at determining when the first AI had *not* picked a good stock, what percent of the time would the AI combination pick a profitable stock?

Further reading

- *The Organization of Behavior*, by Donald Hebb, Wiley.

- *Computing Machinery and Intelligence*, by Alan M. Turing in the journal *Mind*. Vol. LIX (238).

- *Feet of Clay*, by Terry Pratchett, published by HarperCollins, London 2009. This book discusses the fictional concept of golems, which are clay creatures that are *programmed* by a set of instructions written on paper and put into their heads, an interesting analog to robots.

- *A Theory of Human Motivation*, by A.H. Maslow in the journal *Psychological Review*, vol. 50 (4).

- US Dept. of Transportation. *National Motor Vehicle Crash Causation Survey*. `https://crashstats.nhtsa.dot.gov/Api/Public/ViewPublication/811059`.

- *Attacking Faulty Reasoning: A Practical Guide to Fallacy- Free Arguments*, by T. Edward Damer, Cengage Learning.

- *Modern Generative AI with ChatGPT and OpenAI Models*, by Valentina Alito, Packt Publishing.

Answers

Chapter 1

1. What does the acronym PID stand for? Is this considered an AI software method?

   PID stands for **Proportional, Integral, Derivative**, and is a type of closed-loop controller that does not require a model (simulation) to operate. PID is not an AI method because there is no learning or adaptation involved in the decision-making. PIDs are very useful control techniques and are widely used to control motors and thermostats.

2. What is the Turing Test? Do you feel that this is a valid method of assessing an artificial intelligence system?

   The **Turing Test**, originally called *The Imitation Game* by Alan Turing, is an imaginary test, or thought experiment, in which a person communicates with someone or something via a **teletype** (or a text message, for you younger people). An AI would pass the Turing Test if the person was unable to tell whether the entity they were communicating with was a human or a robot. The Turing Test has been passed by modern AI-based chatbots and generative AI engines such as ChatGPT, and new intelligence tests are being created (`https://www.nature.com/articles/d41586-023-02361-7`).

3. Why do you think robots have a problem in general with negative obstacles such as stairs and potholes?

   It is difficult to see negative obstacles (holes, drop-offs, stairs going down, etc.) using the robot's sensors, which have an easier time with positive (going up) obstacles. Usually, cameras and LiDAR can only see part of a negative obstacle due to the bottom being obscured (not visible). It is sometimes easier to reason about negative obstacles by seeing their shadow – the area that the sensor cannot see. The following diagram shows how a robot perceives a hole:

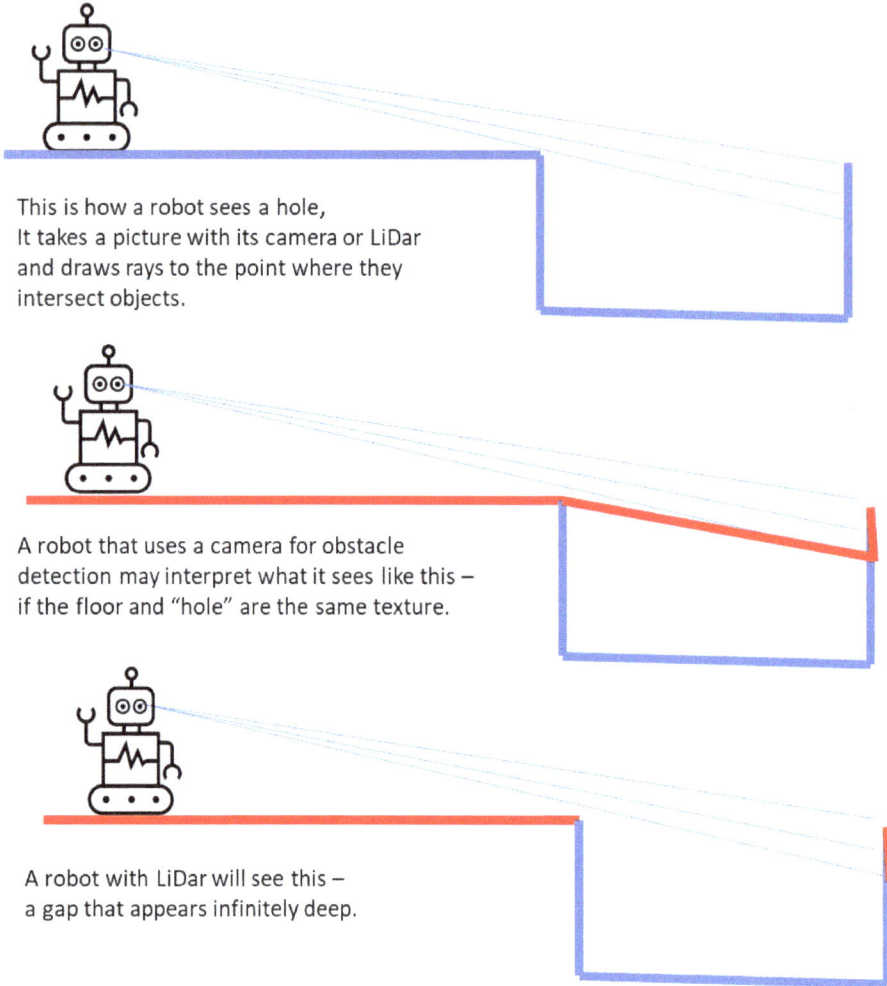

This is how a robot sees a hole,
It takes a picture with its camera or LiDar
and draws rays to the point where they
intersect objects.

A robot that uses a camera for obstacle
detection may interpret what it sees like this –
if the floor and "hole" are the same texture.

A robot with LiDar will see this –
a gap that appears infinitely deep.

Figure 11.1 – How a robot perceives a hole or negative obstacle

The following diagram shows how a robot perceives stairs:

This is how a robot sees stairs. There is nothing solid
to detect. The range of the camera or LiDar is too short.

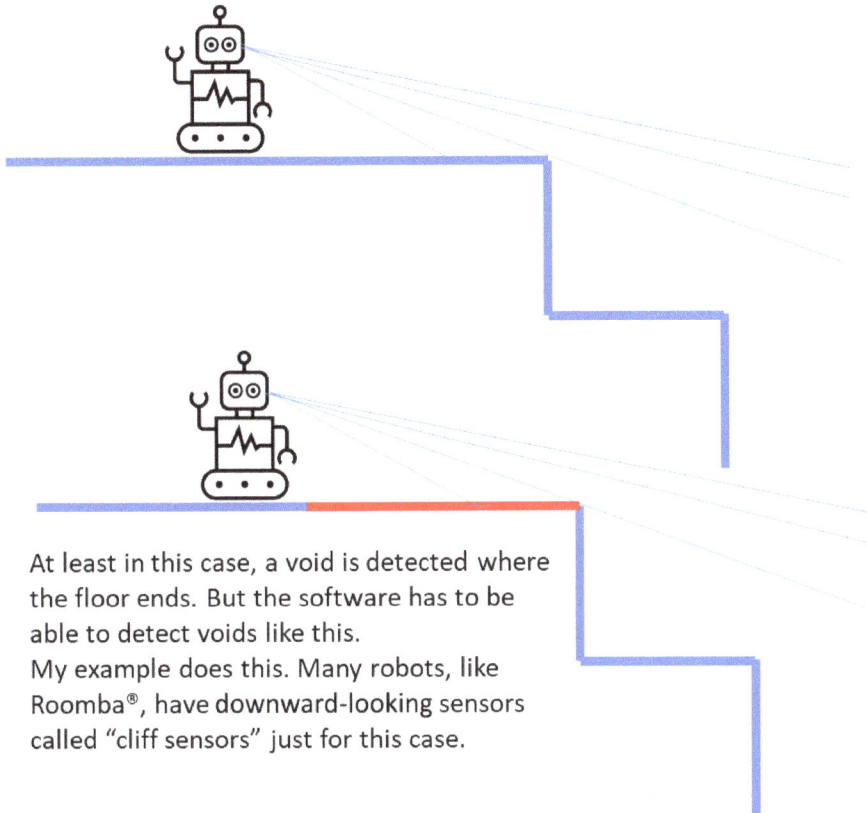

At least in this case, a void is detected where
the floor ends. But the software has to be
able to detect voids like this.
My example does this. Many robots, like
Roomba®, have downward-looking sensors
called "cliff sensors" just for this case.

Figure 11.2 – How a robot sees stairs

4. In the OODA loop, what does the orient step do?

    In the *orient* step, all of the data is put into the same reference frame, which is usually the robot's central point of view. This allows the robot to determine which data is relevant to decision-making. It is the most important step in decision-making.

5. From the discussion of Python advantages, compute the following: You have a program that needs 50 changes tested. Assuming each change requires one run and a recompile step to test. A Make compile and build in C takes 450 seconds and a Python `run` command takes three seconds. How much time do you sit idle waiting for the C compiler?

    Using Python as an interpreted language can save a lot of time on very complex builds, where a C/C++ compiler and link can take 20 minutes or more. The C program test cycle in question would take 6.25 hours to complete, while the Python test program would take 2.5 minutes.

6. What does RTOS stand for?

    RTOS stands for **real-time operating system**. This is an operating system that enforces time limits and processing partitions in the OS itself. An RTOS is a deterministic operating system that always executes a task in the exact same amount of time.

7. Your robot has the following scheduled tasks: telemetry at 10 Hz, GPS at 5 Hz, inertial measurements at 50 Hz, and motor control at 20 Hz. At what frequency would you schedule the base task, and what intervals would you use for the slower tasks (i.e., 10 Hz base, motors every three frames, telemetry every two frames, etc.)?

    You need a number that all of the rates (10, 5, 50, 20) divide into evenly. The smallest number that fits is 100 Hz. I would also accept 50 Hz if the student assumed that the 20 Hz would update two times in one frame and three times in the next frame, which is cheating a little but a common adaptation for a real-time system.

8. Given that a frame rate scheduler has the fastest task at 20 frames per second, how would you schedule a task that needs to run at 7 frames per second? How about one that runs at 3.5 frames per second?

    As given in the previous questions, there does not have to be the same number of samples in each frame in order to come out with a constant frame rate as long as there is a multiple of the base frame rate that every sample divides into. In this case, *20 x 7 = 140*, so the 7 Hz can run at a 20 Hz base rate, and it will repeat patterns every 140 frames, or 7 seconds. Half of 7 is 3.5 and can run at the same base rate with a pattern that repeats every 70 frames, or 3.5 seconds.

    Each update would be 5.7 frames apart, which gets rounded up to 6.

9. What is a blocking call function? Why is it bad to use blocking calls in a real-time system such as a robot?

    A **blocking call** suspends the execution of your program until an interruption or event occurs, such as receiving a datagram or UDP packet. These are bad because you lose control of your program timing and cannot maintain a soft real-time execution. Use **polling calls** instead for serial ports and network interfaces. A polling call looks for data on the interface and then continues when no data is available.

Chapter 2

1. Name three types of robot sensors.

   A sensor is anything that conveys data from the outside world to the robot. Sensors mentioned in the text include the following:

   - Sonar sensors

   - Cameras

   - Microphone

   - Buttons

   - Analog-to-digital voltage sensors

   - Temperature via thermistors

2. What does the acronym PWM stand for?

   PWM stands for **pulse width modulation**, a type of digital-to-analog control scheme where pulses are sent out that get longer based on the amount of control desired. In other words, the pulse duty cycle (amount of time on/off) gets converted into a voltage to drive a motor. This is commonly used to control DC motors.

3. What is analog-to-digital conversion? What goes in and what comes out?

   As the name says, **analog-to-digital** (**A2D**) conversion takes in an analog value, typically a voltage, and converts it into a digital value or number that the digital part of the computer can understand. A typical application is measuring battery voltage to determine the state of charge.

4. Who invented the subsumption architecture?

   As stated in *Cambrian Intelligence: The Early History of the New AI* by Rodney Brooks, the subsumption architecture was originally described by Dr. Rodney Brooks, a professor at MIT who would later help found iRobot Corporation and invent the Baxter Robot. Rodney was trying to develop analogs of insect brains to understand how to program intelligent robots.

5. Compare my diagram of the three-layer subsumption architecture to the three laws of robotics postulated by Isaac Asimov. Is there a correlation? Why or why not?

   No, not really. The three laws of robotics from Isaac Asimov are fictional, while the **Subsumption Architecture** (**SA**) is a real architecture that is used to make robots in the real world.

   Asimov's three laws:

   - Robots will not harm a human being, or through inaction, allow a human to come to harm

   - Robots will obey orders from humans, except when that violates the first law

   - Robots will protect themselves from harm, except when that violates the first two laws

Let's look at the three layers in the SA:

- The bottom layer of the SA is the part that looks inside the robot and takes care of internal systems – I like to compare it to the autonomic nervous system. That protects the robot.

- The second layer is the short-term manager – it tells the robot where to go, which includes obeying orders from users.

- The top layer contains the strategic thinking and planning processes. The correlation is weak, to be truthful.

And, readers, remember the final, or zeroth, law: A robot shall not harm humanity or allow humanity to come to harm. That was a later addition.

6. Do you think I should have given our robot project *Albert* a name? Do you name your robots individually or by model?

Roombas, being robotic vacuum cleaners, exhibit characteristics that people often associate with living entities, such as movement and the ability to navigate spaces autonomously. This behavior can trigger a human tendency to anthropomorphize or attribute human-like qualities to non-human entities. Naming is a natural extension of this anthropomorphism.

7. What is the importance of the ROS_ROOT environment variable?

The most important variables are ROS_ROOT and ROS_PACKAGE_PATH. These variables are used to define the filesystem paths for ROS packages and resources. They are essential for the ROS system to locate and use various packages and resources correctly.

Chapter 3

1. Describe some of the differences between a storyboard for a movie or cartoon and a storyboard for a software program.

A storyboard for a movie is used not only for advancing the plot, but also for showing what point of view will be used – in other words, it is used to plan camera angles, directions, and movements. In that the purpose of both storyboards is to "tell the story" of what happens, they are the same. The point of view of a computer software storyboard should be the user.

2. What are the five *W* questions? Can you think of any more questions that would be relevant in examining a use case?

Who, What, When, Where, Why (with *Why* being the most important). More relevant questions might be: How well? How often? How many or how much?

3. Complete this sentence: A use case shows what the robot does but not _____.

"How it does it." Use cases are from the user's perspective and never include implementation details.

4. Take the storyboard in *step 9*, where the robot is driving to the toybox, and break it down into more sequenced steps in your own storyboard. Think about everything that must happen between *frames 9* and *10*.

    The robot has to do the following:

    I.   Determine a route to the toybox.

    II.   Plan a path.

    III.   Avoid obstacles along the way.

    IV.   Align itself with the front of the toybox.

    V.   Drive up to the toybox.

    VI.   Move the robot arm to clear the top.

5. Complete the reply form of the "knock-knock" joke, where the robot answers the user telling the joke. What do you think is the last step?

    It is to compliment the teller of the joke – the robot should say "That is very funny" or "I am sorry, I am unable to groan". Yes, that is my opinion and not an official joke writer's idea. What do you think?

6. Look at the teleoperate operations. Would you add any more or does this look like a good list?

    The robot needs to send video back to the operator so that the operator can see where they are going.

7. Write a specification for a sensor that uses distance measurement to prevent the robot from driving downstairs.

    The robot shall have a sensor capable of detecting negative obstacles in the floor (i.e., stairs going downward, balconies) at a distance of at least six inches from the robot along the robot's driving direction.

8. At which distance can a camera with 320x200 pixels and a 30-degree **field of view** (**FOV**) vertically and horizontally see a 6" stuffed animal, assuming we need 35 pixels for recognition?

    To solve this problem, we first need to determine how many degrees per pixel we have and then use that to calculate the angular dimension of the target, which is 35 pixels tall:

    30 degrees / 320 pixels wide = 0.0937 deg/pixel

    To find the number of degrees per pixel, we can perform the following calculation:

    35 pixels* deg/pixel = 3.28 degrees

    This gives us an isosceles triangle, but we need a right triangle to do the math. Divide the base into two to make a right triangle; thus, the base of this triangle is now *3 inches*.

We also divide the angle in half:

3.28/2 = 1.64 degrees

Then, to calculate the perpendicular height, we divide the length of the base by the value of *tan*:

3 / tan(1.64) = 104.78 inches

This translates to 8.73 feet.

This can be illustrated by the following diagram:

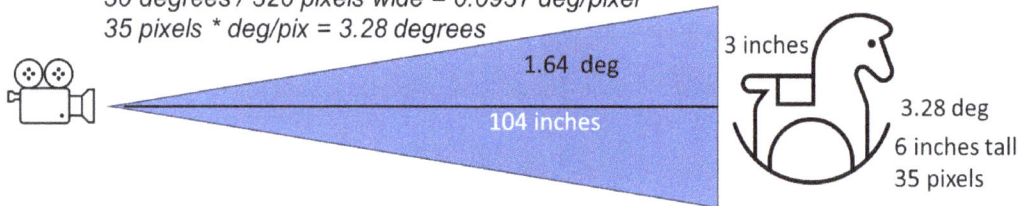

Figure 11.3 – Solving for pixels needed for recognition

Thus, the required distance is 8.73 feet.

Chapter 4

1. We went through a lot in this chapter. You can use the framework provided to investigate the properties of neural networks. Make adjustments to the learning rate, batch size, number of epochs, and loss functions.

   This is an exercise for the student. You should see different curves develop as these parameters are changed. Some will not produce an answer at all (which looks like random results – the curve stays at the same level as no learning is taking place). Some will learn faster or slower.

2. Draw a diagram of an artificial neuron and label the parts. Look up a natural, human biological neuron and compare.

   See *Figure 4.3* in the chapter. The artificial neuron has a number of inputs, a set of weights, one for each input, a bias, an activation, and a set of outputs.

3. Which features are the same in a real neuron and an artificial neuron?

   Both have multiple inputs and multiple outputs and accept inputs, perform some processing, and then make an output. Both use some sort of activation function (the biological equivalent is the synapse) to determine when to *fire* or produce an output. Both are part of networks: it takes a lot of neurons to compose a neural network, and likewise an animal brain.

4. Which features of a real neuron and an artificial neuron are different?

The natural **neuron** is an analog device that can handle many levels or degrees of input, with no simple on/off binary representations like the computer neuron. Neurons use chemical paths that make pathways and connections easier the more they are used, which is the learning function of a neuron. This is simulated by the weights in an artificial neuron. The natural neuron has an axon, or connecting body, that extends out to the outputs that can be at a quite distance from the nerve inputs. Neurons are randomly connected to other neurons, while artificial neurons are connected in regular patterns. Some neural networks use **dropout layers** that randomly disconnect neurons, providing some randomness to the output that helps the network estimate non-linear solutions.

5. What relationship does the first layer of a neural network have to the input?

The first layer contains the number of inputs to the network. For instance, if you have five inputs, then the first layer must contain five neurons.

6. What relationship does the last layer of a neural network have to the output?

The last layer of an ANN is the output layer and has to have the same number of neurons as there are potential outputs.

7. Look up three kinds of loss functions and describe how they work. Include mean square loss and the two kinds of cross-entropy loss.

Loss functions in ANNs are the error functions that compare the expected output of the neuron with the actual output. Let's look at them in detail:

- **Mean square loss** (MSL): This is the most commonly used loss function. It is given by the sum of the squares of the distances between the output and the expected output. MSL amplifies the error the farther away from the desired solution it is.

- **Cross entropy** (XE): This is also called log loss and is used mostly for the classification of CNNs. As the predicted value approaches 1 (no error), XE slowly decreases. As the values diverge, XE increases rapidly. Two types of cross entropy are as follows:

  - Binary (on/off, used for yes/no questions)

  - Sigmoid cross-entropy, which can handle multiple classes

8. What would you change if your network trained to 40% and got "stuck" or was unable to learn anything further?

You are probably *overfitting* and have too small a sample size or your network is not wide or deep enough.

Chapter 5

1.   In Q-learning, what does the Q stand for (you will have to research this on the internet)?

The origin of **Q-learning** is the doctoral thesis of Christopher John Cornish Hellaby Watkins from King's College, London, May, 1989 (`https://www.researchgate.net/publication/33784417_Learning_From_Delayed_Rewards`). Evidently, the Q just stands for *Quantity*.

2.   What could we do to limit the number of states that the Q-learning algorithm must search through?

Only pick the Q-states that are relevant and are follow-ons to the current state. If one of the states is impossible to reach from the current position, or state, then don't consider it.

3.   What effect does changing the learning rate have on the learning process?

If the learning rate is too small, the training can take a very long time. If the learning rate is too large, the system does not learn a path but instead overshoots and may miss the minimum or optimum solution. If the learning rate is too big, the solution may not converge or may suddenly drop off.

4.   What function or parameter serves to penalize longer paths in the Q-learning equation? What effect does increasing or decreasing this function have?

The discount factor works by decreasing the reward as the path length gets longer. It is usually a value just short of 1.0, for example, 0.93. Making the discount factor higher may cause the system to reject valid longer paths and not find a solution. If the discount is too small, then the paths may be very long.

5.   In the genetic algorithm, how would you go about penalizing longer paths so that shorter paths (fewer number of steps) would be preferred?

You would adjust the fitness function to consider path length as a factor in the fitness calculation.

6.   What effect does changing the learning rate in the genetic algorithm change? What are the upper and lower bounds of the learning rate?

Generally, increasing the learning rate shortens the learning time in generations, up to a limit where the path jumps out of the valid range. For our example program, the lowest learning rate that returns a valid solution is five, and the highest value is 15.

7.   In the genetic algorithm, what effect does lowering the population cause?

It causes the simulation to run much faster but take many more generations to find a solution.

Chapter 6

1. Do some internet research on why an open source voice assistant was named Mycroft. How many stories did you find and which one did you like?

   I found at least three. My favorite is that Mycroft is Sherlock Holmes' older, and some say smarter, brother. Sherlock Holmes is played on TV in the UK by Benedict Cumberbatch, who also played Alan Turing in the movie *The Imitation Game*, the original name of the Turing Test, a test of AI conversation, which is what Mycroft does.

2. In the discussion of intent, how would you design a neural network to predict command intent from natural language sentences?

   One approach would be to gather a selection of commands, label the intent of the command, use the commands as input in a neural network, and use intent as the output label for the training.

3. Rewrite the "receive knock-knock jokes" program to remember the jokes told to the robot by adding them to the joke database used by the "tell knock-knock jokes" program. Is this machine learning?

   It is fairly simple to add a program to just write to the knock-knock joke program database. You'll find a version of this in the GitHub Repository. Is this machine learning? I would say definitely! The machine has a capability that it did not have before. It did not have to be reprogrammed to have the new capability, so it learned.

4. Modify KnockKnock (the program that tells knock-knock jokes) to play sounds from a WAV file, such as a music clip, as well as do text-to-speech.

   Add these lines to the KnockKnock program:

   ```
   "play_wav_cmdline": "paplay %1 --stream-name=mycroft-voice",
   ```

   You can use this to play the audio. You can also add a tag to the joke file that indicates a WAV file <groan.wav>. Then, if you see this tag, call the play_wav_cmdline function above.

5. The sentence structure used in this chapter is all based on English grammar. Other languages, such as French and Japanese, have different structures. How does that change the parsing of sentences? Would the program we wrote be able to understand Yoda?

   In other languages, the object or the subject appears in a different order, just as in Yoda's speech patterns. "Backwards, I talk," Yoda would say. This does require us to change or add new sentence patterns to our .voc files.

   You can follow Mycroft's instructions for changing the engine to understand French at https://mycroft-ai.gitbook.io/docs/using-mycroft-ai/customizations/languages.

6. Do you think that Mycroft's Intent Engine is actually understanding intent or just pulling out keywords?

I do not place Mycroft in the category of construction AI chatbots, but it is rather a referential type that looks up answers in a database, which makes it more of an expert system than an AI program. It does use AI neural networks in the speech-to-text section.

7. Describe the voice commands necessary to instruct the robot to drive to an object and pick it up without the robot being able to recognize the object. How many commands do you need?

We need two commands:

- *Can you see any objects?*

- *Drive to the closest object*

8. From *Question 7*, work to minimize the number of commands. How many can you eliminate or combine?

The following are the voice commands to tell the robot to drive to the nearest object:

- *Hey, Albert*

- *Drive to the nearest object*

9. From *Question 7*, how many unique keywords are involved? How many non-unique keywords?

The four keywords are **see**, **objects**, **drive**, and **closest**.

All of the words are unique except **objects**, so there are three unique words. They have not been otherwise defined in the Mycroft word database.

Chapter 7

1. Regarding SLAM, what sensor is most used to create the data that SLAM needs to make a map?

Light detection and ranging (**LiDAR**) sensors are the most common SLAM sensors used by a wide margin. The 3D data that LiDAR provides is perfect for SLAM's mapping function.

2. Why does SLAM work better with wheel odometer data available?

The wheel odometers reduce the search space that the SLAM algorithm needs to look for the possible locations of the robot after moving. Thus, it increases information and reduces uncertainty in the map. How does it do this? By giving extra measurements about where the robot is located (how far it moved), we can then reduce our search to match where we are against our sensor readings.

3. In the Floor Finder algorithm, what does the Gaussian blur function do to improve the results?

The **Gaussian blur function** reduces noise and gets rid of stray single pixels in the image, making for a smoother result.

4. The final step in Floor Finder is to trace upwards from the robot's position to the first red pixel. In what other way can this step be accomplished (referring to *Figure 7.3*)?

    Instead of using radial red lines, the program can just draw upwards from the bottom of the screen in a series of vertical lines.

5. Why did we cut the image in half horizontally before doing our neural network processing?

    We just want to use the upper half of the room to train the network because the lower half has the toys on it and is subject to change. The upper half of the room does not change with the addition of toys.

6. What advantages does using the neural network approach provide that a technique such as SLAM does not?

    We don't have to have a map to successfully navigate the room. We are providing labeling of our training set by just driving the robot around and taking pictures at regular intervals. This approach is also far more resilient to changes in the room, such as the furniture being in slightly different positions. Please also see `https://hackaday.com/2021/10/25/fast-indoor-robot-watches-ceiling-lights-instead-of-the-road/` for someone else's implementation of this idea, used for indoor car racing.

7. If we used just a random driving function (where you make random turns at random times) instead of the neural network, what new program or function would we have to add to the robot to achieve the same results?

    We would need to have a navigation function that determined where in the room we were at – this would probably mean a SLAM algorithm. We would also need something to detect the stairs.

8. How did we end up avoiding the stairs in the approach presented in the chapter? Do you feel that this is adequate? Would you suggest any other means to accomplish this task?

    We trained the robot to navigate by looking at the upper part of the room. We only drove the robot in safe areas and used that information to allow the robot to predict its next driving command based on where it was in the room. Since we did not drive the robot down the stairs in this process, the robot will never get a command to drive toward the stairs. If there is a toy near the stairs, the robot will still go pick it up but will drive away from the stairs afterward when it goes back to navigation mode. We have to be careful to get a good training result before letting the robot loose, however. I used a baby gate to block the stairs for early testing. As an additional safety measure, we can add a lookdown sensor to detect stairs. I would use an **infrared proximity detector** (**IRPD**) for this purpose.

Chapter 8

1. What are the three ways to traverse a decision tree?

    From beginning to end (start to goal); from goal to start; and from both ends at once to meet in the middle.

2. In the fishbone diagram example, how do you go about pruning the branches of the decision tree?

    By eliminating the effect of the item on a branch. For example, using our "robot does not move" fault, if the branch says "Arduino-no power" and you check to see if the Arduino has power and it does, you can prune that branch. If the branch is "motor stuck", the effect of having a motor stuck is that the robot will drive in circles. As the robot is not driving in circles – it is not driving at all – you can prune that branch.

3. What is the role of the Gini coefficient in creating a classification?

    It determines the amount of *impurity* in the sample or pool. When the Gini coefficient = 0, all of the members of the class have the same attributes, and no further subdivision is possible. This minimizes misclassification. The Gini coefficient is given by one minus the sum of the square of the probability of an item being in that class.

4. In the toy classifier example using the Gini coefficient, which attributes of the toy were not used by the decision tree? Why not?

    Color, Noise, Soft, and Material were not useful for dividing the categories by labels as the labels and the items did not correlate. It does make sense that color is not useful for dividing toys by type.

5. Which color for the toys was used as a criterion by one of the classification techniques we tried?

    The color white was used by the decision tree that used the Gini index and one hot encoding to separate the stuffed animals.

6. Give an example of label encoding and one hot encoding for menu items at a restaurant.

    Let's have three types of menu items: appetizer, entrée, and dessert. Label encoding would substitute 0 for appetizer, 1 for entrée, and 2 for dessert. One hot encoding would use 1 0 0 for appetizer, 0 1 0 for entrée, and 0 0 1 for dessert.

7. In the A* algorithm, discuss the different ways that G() and H() are computed.

    The G() function is the distance along the path from the current position to the start. H() is the distance from the current position directly to the goal (the Euclidean distance). Note that G() follows the path and H() is the straight-line distance to the goal since we have not computed a path to the goal yet.

8. In the A* algorithm, why is H() considered heuristic and G() not? In the D* algorithm, heuristics are not used. Why not?

    Heuristic is an approach that is not guaranteed to be optimal but instead is sufficient for the task. Since H() – the direct line distance to the goal – is an estimate and ignores any obstacles, it can't be used directly but is just a way to compare one position to another. A major difference between D* and A* is that D* starts at the goal and works backward toward the start. This allows D* to know the exact cost to the target – it is using the actual path distance to the goal from the current position and not a heuristic approach or an estimate of the distance to go, as A* did.

9. In the D* algorithm, why is there a RAISED and a LOWERED tag and not just a CHANGED flag?

    The RAISED squares or points are eliminated from consideration. The LOWERED squares may be added back into the queue for consideration to be a path. Keep in mind that lowering scores due to new sensor readings ripples through the path planner.

Chapter 9

1. What is your favorite movie robot? How would you describe its personality?

    This is, of course, a subjective question. I'm a big R2D2 fan. R2 is feisty, determined, and stubborn as well as being a faithful companion and helper. R2 will get you out of a jam, fix your starfighter, provide cover from hostile fire, and hack Imperial computers. He is a Swiss army knife on wheels.

2. What techniques did the movie-makers use to express R2D2's personality (body language, sounds, etc.)?

    R2D2 owes his personality to a combination of his emotional beeps and squawks (provided by Ben Burtt) and his body movements provided by having a person inside his chassis (Kenny Baker). They were stuck with the not-very versatile chassis designed for the first *Star Wars* movie, which only has a head that moves. Most of R2's persona comes through in his sounds, including his famous scream.

3. What are the two types of chatbots? List some of the strengths and weaknesses of each.

    The two types of chatbots are as follows:

    - **Retrieval-based chatbots**: Retrieval-based chatbots look up responses in lists of scripts and choose from a number of phrases that are written in advance by humans. The strengths of these chatbots are that they are easy to program, allow more control over the outputs, and are much smaller and faster programs. The weaknesses are that they have limited responses and the use of keywords gives them a small vocabulary.

- **Generative chatbots**: Generative chatbots use the rules of grammar and models of sentences to create new sentences with proper meaning, are more flexible, and can handle a wider range of topics, but they are much harder to program and are complex and slow (comparatively speaking). Generative chatbots have now taken over, given the success of ChatGPT and other generative AI models.

4. In *Figure 9.2*, the illustration on modeling custom distributions (the airport example), the lower picture shows two standard distributions and two uniform distributions. Why don't the curves go all the way to the top of the graph?

    The two distributions will add together – the standard distributions sit *on top* of the uniform distributions, and the two combined go to the top of the graph.

5. Design your own robot emotions: Pick six contrasting emotions that can express the entire range of your robot's personality. Why did you pick those?

    This is another subjective question. My answers are in the text. I picked emotions that represented the range of capability of my robot and the situations it would be in. I kept to a friendly type of robot so that the only negative emotion was sadness – there was no anger, for instance.

6. If you were designing a robot to have the personality of an annoying little boy (think Bart Simpson, Cartman, or Dennis the Menace if you are that old), what traits would it have?

    A small boy would be mischievous, have a short attention span, constantly change the subject, keep trying to bring up the same topic over and over, and repeat variations of the same questions. How might we represent mischievous? Perhaps by ignoring directives and generating random events or distractions that get the robot's attention away from the task at hand.

7. Why is it important for the robot to have a backstory or biography?

    To provide consistent answers to personal questions, such as "How old are you?"

8. For the following questions, pick a persona from my list to model (from *Integrating Artificial Personality*).

    I.   Write six lines of dialogue for the robot to ask a human where they last went on vacation.

    - So, where did you go on vacation last?

    - Summertime is coming up. Where did you go on vacation last year? Do you like to travel? Where have you been?

    - I never get to go on vacation. Where did you go last?

    - I have heard of this concept called vacation. Where do you like to go? Have you been to the beach?

II.   Write six ways for the robot to express that it is tired and needs to recharge without sounding like a robot.

- I'm tired – have you seen my recharger?

- Wow, it is getting late. I've been at this for a long time.

- Well, my battery is getting low. Must be about quitting time. I am starting to feel a bit run down.

- Well, look at the time! My battery needs attending to. I'm getting hungry in here. Can I go charge now?

Chapter 10

1.   Given that we started the chapter with knock-knock jokes and ended up talking about robot phobia and addressing philosophical questions about existence, do you feel that AI is a threat? Why or why not?

I do not feel that robots or AI are a threat in any way because the necessary and sufficient conditions for robots to be a threat do not exist, which is to say that the robots have to *want* to take over the world and must have a *need* to take over. Currently, robots and AI have no such wants or needs.

2.   List any five professions that would be necessary to turn our Albert robot into a product.

We would need project managers, packaging designers, advertising and marketing experts, salespeople, engineers, technicians, artists, package designers, machinists, electricians, accountants, lawyers, a psychologist, and support staff, among others. You can select any five from among these options.

3.   Why would we need a psychologist in our imaginary company that manufactures robots?

Psychologists study normal and abnormal mental states and cognitive processes, which is exactly what we are trying to simulate in an artificial personality. We want the robot to not trigger bad responses in people. I once had a robot with flashing red eyes that caused small children to have panic attacks. Psychologists would help avoid such errors.

4.   What components found in cell phones or smartphones are also found in quadcopters?

GPS receivers, radios, Wi-Fi, Bluetooth, accelerometers, gyroscopes, and, these days, applications, or apps.

5. Why are artificial intelligence systems, specifically artificial neural networks, naturally non-deterministic in both result and time?

    They are universal approximation systems that work in probabilities and averages, not in discrete numbers and logic. Artificial neural networks can take a different amount of time because a particular bit of data may take different paths at different times, going through a different number of neurons and thus not taking the same amount of time to process. It is true that if you provide the exact same input to a neural network, it will give you the exact same answer every time. In the real world that robots live in, however, the case that two inputs are identical in every way can be very rare.

6. What might be a practical application of an AI system that made predictable mistakes?

    You can use a neural network-based system to model a bad human operator for a driving simulation to help teach other drivers (and self-driving cars) how to avoid bad drivers. The desired state is an unpredictable driver, so just train the neural network to 60% or so. Now the network will come up with the wrong answer 40% of the time, i.e., be statistically predictable. I actually did this on a project for the Navy that wanted a simulation of imperfect people misusing a system at a predictable level, so they could create a responsive control system that could handle those mistakes.

7. If an AI system was picking stocks for you and predicted winning stock 43% of the time and you had a second AI system that was 80% accurate at determining when the first AI had *not* picked good stock, what percent of the time would the AI combination pick profitable stock?

    We have 100 stocks picked by our AI program. Of that set, an indeterminate number are winners and losers. There is a 43% chance the stock is correctly predicted to be a winner and a 57% chance it is predictably a loser. We have no way of judging the stocks as being winners or losers except by investing our money, which is what we are trying to avoid – investing in bad stocks. A 43% chance of winning is not good.

    The second AI has an 80% chance of telling you that the first AI chose bad stock. Eighty times out of 100, you will know that the stock was not a winner. You are left with an 80% chance of correctly identifying one of the 57 bad stocks, which eliminates 45 stocks. That leaves you with 55 stocks, of which 43 are winners (on average), which raises your odds to 78%.

    Bayes' theorem shows the combination of two independent probabilities (probability of x occurring given c has occurred):

    px = probability of x, and pc = probability of c)

    $$p(x|c) = (px^* pc) / (px^*pc)+(1-px)(1-pc)$$

    Using this theorem, I recomputed the combined probabilities as 75.1%, so I'll take either answer.

Appendix

Robotic Operating System (**ROS**) was a framework designed to enable the development of software for complex robots and was developed by a company called *Willow Garage*, specifically for the control of the PR2 robot. The PR2 was a human-sized robot with two **7-degree of freedom** (**7DOF**) arms and an entire array of sensors. Controlling this very complex robot required the interaction of a multitude of sensors, motors, and communications. The ROS framework allowed the development of robot components to be done independently. While not an operation system in the traditional sense of the word, it is a **Modular Open Source Architecture** (**MOSA**).

The primary tool of ROS is a robust *publish-subscribe* service that makes talking between processes — that is, **Inter-Process Communications** (**IPC**) — easy and flexible. It also standardized a lot of the interfaces between sensors, motors, and controls for robots.

We will be using ROS 2 throughout this book. ROS 2 is a new version of ROS. In this *Appendix*, we will discuss how to install ROS 2, use it to communicate, and briefly introduce some of the tools ROS provides. We will provide an in-depth introduction to ROS 2 and describe some of the hardware involved in the design of our toy-collecting robot, Albert. We will also cover some of the hardware referenced in the book when creating Albert as our example for this book. Albert V2, the robot in this second edition of this book, is my 30th or so robot design.

The topics covered in the appendix include the following:

- Introducing MOSA
- A brief overview of ROS 2
- Software requirements for the robot
- Introducing the hardware for the robot
- Robot safety tips

Introducing MOSA

ROS is an example of a MOSA. Why is this important? Imagine if every electrical appliance in your house had its own plug, a different voltage, and a different wire. It would make life very difficult for you. But all your electrical plugs are the same shape and put out the same voltage. They are standardized interfaces that allow you to plug many different types of appliances into them. A MOSA acts like that for software, standardizing interfaces and allowing *plug-and-play* compatibility.

The following are its advantages:

- A MOSA system architecture allows modularity – the ability to create software in sections or modules that can be developed, debugged, and operated independently. Before ROS, I created one major executable that ran everything on my robot. The problem with this is, first of all, that I could not take advantage of the multi-core nature of my **Single Board Computer** (**SBC**), which was the robot's brain. I had all my code in one thread, in one program, and splitting out functions to run independently was difficult.

- Then there were the interactions. If I changed the timing on my motor driver, it messed up the sensor timing for the camera. If I changed the path planner, then the steering needed adjusting. This sort of interaction is typical in a **unitary architecture**. However, in a MOSA, each section of the robot, for example, the robot arm controller, is independent and runs in its own program. They can be developed and debugged independently, and interactions are limited to the interfaces we create in those programs. This frees us from a lot of problems we would otherwise have.

- The other feature we can take advantage of is the very large library of standard, already-created interfaces and programs that ROS provides. We don't need to create a steering interface; ROS has one (the `Twist` command). We don't need to create data types for camera imagery data; ROS has several to choose from. ROS also has USB camera drivers and viewers that we can use without writing any code at all.

Now let's talk a bit about how ROS 2 works.

A brief overview of ROS 2

As mentioned earlier, ROS 2 is the latest version of ROS, a widely used framework for developing robot applications. I've been using ROS for some time now and appreciate how much simpler it makes the integration of various components, sensors, and capabilities into my robots. I resisted moving to ROS for some time, but now that I have invested the time to learn what it can do, I can't imagine developing a robot without it.

In this section, we will discuss some concepts that are foundational for our understanding of ROS 2, how to install ROS 2, and some basic commands that we can use with it.

> **Note**
> Please use whatever is the latest version of ROS 2.

Understanding the basic concepts

ROS 2 works a bit differently than other programming paradigms. ROS is based on a publish/subscribe mechanism that allows different programs or processes to pass information from one to another without having to know in advance who is receiving the data. Let's look at how this process works:

- Each program or code that communicates in ROS is called a **node**. Nodes each have names that uniquely identify them to the system.

- Nodes publish data on **topics**, which represent an interface for messages.

- **Messages** are interfaces that have data types (string, float, fixed, array, etc.).

For example, a joystick interface node publishes to the commanded velocity (`cmd_vel`) topic with a `Twist` message. On the receiver end of the message, the motor control interface subscribes to the `cmd_vel` topic to receive Twist messages that have speed and turn information for the robot.

> **Note**
>
> Why is it called a Twist message? The reason for using the term "twist" is conceptual. Think of a robot's movement as a combination of straightforward (linear) motion and rotation (twisting or turning). By combining linear and angular velocities, a Twist message effectively describes how the robot "twists" through space, which includes both translation (moving from one place to another) and rotation (changing orientation).

ROS has a few other useful features:

- There is a systemwide **logging** functionality with multiple levels of information. Log messages can be labeled as DEBUG, INFO, WARN, ERROR, or FATAL. Log messages are automatically displayed on the local output (usually the command line). They are collected centrally, and you can parse through log messages to debug problems. Periodically, you do need to purge old log files, as they do tend to pile up. Logs are put into the `~/.log` directory.

- One of my personal favorite features is **parameters**. These are external data values that can be created independently of source code and can be used to turn features on or off, or to set critical settings, such as image size, resolution, range, and other features. The ability to have external parameters goes a long way to make ROS portable and the interfaces reusable. Parameters can be specified in launch files, which are ways of starting multiple programs (nodes) all at once. Albert has five major subsystems (control, motor drive, arm control, vision, and speech), all of which have to be started together.

Comparing ROS 2 and ROS

ROS 2 introduced some very significant improvements over the original ROS. Some of them are as follows:

- The most notable improvement is the absence of the **roscore** *traffic cop* application. This central program acted as the Master node in any ROS implementation and directed the other nodes on where to communicate via sockets. It directed all the traffic on the network. Therefore, every node had to communicate with roscore to know where it was running. If roscore died, or was turned off, then the entire ROS set of applications stopped. Instead, ROS 2 uses **Data Distribution Services** (**DDS**) as its *middleware layer*. DDS is a standard for high-performance, scalable, and real-time data exchange, created by the **Object Management Group** (**OMG**) and managed by the DDS Foundation. It provides a decentralized discovery mechanism for nodes to find each other without needing a central master such as roscore. When a node starts, it advertises its presence to other nodes and also discovers existing nodes and topics. This process is managed by the underlying DDS implementation.

- ROS 2 has more features for **real-time processing**, and does not have the weird workaround that ROS had, where we had to do things such as post-dating messages into the future. These features include using DDS (which is intended for real-time systems, the ability to work with **real-time operating systems** (**RTOSs**), and to use pre-emption to control processes.

- One big improvement is that ROS 2 will **run natively on Windows** for the first time. You don't need virtual machines to run ROS on Windows – you can build your own control panels and interfaces directly in Windows.

- ROS 2 also improves **discovery** (the process of finding nodes on the network) and has some enhancements for cyber security.

So, all in all, ROS 2 was well worth taking the time to upgrade, and I feel that it is far less fiddly than the old ROS, easier to set up on both Windows and Linux, and easier to manage.

Let's look at software requirements next.

Software requirements for the robot

In this section, we will discuss the software requirements for the robot and how to install them on the robot's CPU.

Installing ROS 2

The version of ROS 2 I installed on Albert currently is *Foxy*. Please feel free to use the latest version of ROS 2. Jetson tends to run behind Ubuntu upgrades and is several releases behind. We use the Jetson Nano because it has the required power and the **Graphics Processing Units** (**GPUs**) to run neural network software. My version of the Nano is running Ubuntu 20.04, but you should also be able to get it working with Ubuntu version 18.04.

I used the standard ROS 2 installation script that can be found at `https://github.com/jetsonhacks/installROS2`. This is a script that contains all of the steps found on the regular ROS 2 installation page at `https://docs.ros.org/en/foxy/Installation/Alternatives/Ubuntu-Development-Setup.html`.

Note that we are doing an *Install from Source* setup since many of the programs need to be recompiled to run on the Jetson Nano's ARM-based architecture. For me, this process took about 4 hours, so it does take some patience.

Installing other packages

You will also need to install the following Python packages: **Scientific Python (SciPy)**, **Numeric Python (NumPy)**, **scikit-learn**, **Open Computer Vision (OpenCV)**, and **PyTorch**. Let's look at the commands needed to install these packages:

- SciPy:

```
python -m pip install scipy
```

- NumPy:

```
pip install numpy
```

- scikit-learn:

```
pip3 install -U scikit-learn
```

- OpenCV:

```
sudo apt-get install python3-opencv
```

    For more details on OpenCV, you can refer to `https://docs.opencv.org/3.4/d6/d00/tutorial_py_root.html`

- PyTorch:

```
pip3 install torch torchvision torchaudio --index-url https://
download.pytorch.org/whl/cu118
```

Now let's look at how we can get started with ROS 2.

Basic ROS 2 commands

The following commands relate to starting, controlling, and monitoring nodes in ROS 2:

1. To start executing a package in ROS 2, use the following command:

    ```
    ros2 run <package name> <executable name>
    ```

    For example, if we wanted to run turtlesim, the standard example program provided with ROS 2, then we would use the following command:

    ```
    ros2 run turtlesim turtlesim_node &
    ```

    This would execute the turtlesim_node from the turtlesim package in the background.

2. To check that this node is running, we type the following:

    ```
    ros2 node list
    ```

    This gives us a list of currently running nodes:

    ```
    /turtlesim
    ```

3. We can get more information about the node by typing this command:

    ```
    ros2 node info /turtlesim
    ```

    This results in the following:

    ```
    /my_turtle
     Subscribers:
     /parameter_events: rcl_interfaces/msg/ParameterEvent
     /turtle1/cmd_vel: geometry_msgs/msg/Twist
     Publishers:
     /parameter_events: rcl_interfaces/msg/ParameterEvent
     /rosout: rcl_interfaces/msg/Log
     /turtle1/color_sensor: turtlesim/msg/Color
     /turtle1/pose: turtlesim/msg/Pose
    ```

Here, we can note the following key points:

- The turtlesim publishes parameter_events (which happen when you change parameter values), it has a logging interface to /rosout, and it publishes the color of the turtle robot and its position (pose)

- For input, it subscribes to cmd_vel, which uses the Twist command to move the turtle, and to parameter_events, which allows the program to receive parameter changes

The command for looking at topics is as follows:

```
ros2 topic list
```

This shows topics that are active. For our turtlesim example, we get the following output:

```
/parameter_events
/rosout
/turtle1/cmd_vel
/turtle1/color_sensor
/turtle1/pose
```

If you add -t to the end of that command, you also get the topic message type:

```
/parameter_events [rcl_interfaces/msg/ParameterEvent]
/rosout [rcl_interfaces/msg/Log]
/turtle1/cmd_vel [geometry_msgs/msg/Twist]
/turtle1/color_sensor [turtlesim/msg/Color]
/turtle1/pose [turtlesim/msg/Pose]
```

You can find more details and the full tutorial at https://ros2-industrial-workshop. readthedocs.io/en/latest/_source/navigation/ROS2-Turtlebot.html. Also, I recommend that you refer to the topic list in the tutorials available from the ROS 2 website: https:// docs.ros.org/en/foxy/Tutorials.html. This will give a more robust introduction to ROS 2. The rest of what you need to run the programs in this book is in the text.

Now, let's look at the hardware that makes up Albert the Robot.

Introducing the hardware for the robot

I designed Albert the Robot to perform one manual task – picking up toys. As such, I chose a set of motors, a speaker, and a robot arm as effectors, and a camera and microphone as sensors. Here is a labeled diagram of what Albert looks like:

Figure 12.1 – Albert the Robot

In the following sections, we will look at how I put Albert together.

Effectors – base, motors, and wheels

The robot base is a two-layer aluminum allow frame that I purchased at https://www.amazon.com/gp/product/B093WDD9N5. This base uses **Mecanum** wheels, which have the unique ability to move the chassis not just forward and back, but sideways and at any angle. For video game fans, this sideways movement is sometimes called **strafing**. You will note that the wheels have smaller rollers mounted at 45-degree angles. These convert various inputs into multiple directions. Moving all four motors forward, not surprisingly, results in the platform moving forward. Moving the right wheels forward and the left wheels backward results in turning in place to the right.

When we move the left wheels away from each other and the right wheels towards each other (left front forward, right front backward, left rear backward, and right rear forward), the vehicle moves sideways – strafes – to the right. Reverse these directions and you strafe to the left. These are the motions we need to complete the exercises in the book.

I made some modifications to the base: I cut off the little white connectors to the four motors so I could wire them to the motor controller.

Battery

I have used a 4,200 mAh **Nickel Metal Hydride (NiMH)** battery– no lithium battery here, so there's less risk of fire. The output is 7.2 V. This should provide several hours of runtime to our robot, and it fits in the chassis. You can use any drone-type battery with the same specs. A larger battery won't fit in the chassis, which is quite small. This battery is adequate for our needs. Here is an example of a battery you can get for your robot: `https://www.amazon.com/gp/product/B08KXYY53G`.

DC/DC power supply

One of my favorite bits of kit is this DROK DC/DC adjustable power supply. This board provides the 5 V power to the main computer and the Arduino. You adjust the voltage level to 5 V with a small screwdriver. I really like to have this display on the robot to show that everything is working. Here is a link to the one I used: `https://www.amazon.com/Converter-DROK-Adjustable-Stabilizer-Protective/dp/B01FQH4M82?th=1`.

CPU – the brains of the outfit

As we've mentioned throughout the book, the main computer for Albert is the **Nvidia Jetson Nano**. This is a rugged little single-board computer with a big heat sink. This CPU is specifically designed to run AI code, with 128 **graphics processing units (GPUs)** and four Arm A57 **central processing units (CPUs)** running at 1.43 GHz, and packed with peripheral ports and external I/O capability. This is an ideal board for our needs for running a robot with perception and decision-making abilities. The Jetson sits on a development board, which provides an interface to all of its capabilities. Any of the other members of the Jetson family (TX2, Xavier, and AGX) will also work, but I've sized the code in the book to fit on the Nano specifically. This is a link to purchase the Jetson Nano computer: `https://developer.nvidia.com/embedded/jetson-nano-developer-kit`.

I've added a Wi-Fi card, which is mounted underneath the CPU. This is the one I used: `https://www.amazon.com/gp/product/B07SGDRG34`.

Effectors – robot arm

I can't say enough about how much of an upgrade the digital servo arm used in this book (`https://www.hiwonder.com.cn/store/learn/42.html`) is over the previous analog servo arm I used in the first edition of the book. Some of the advantages of this arm are as follows:

- This arm uses **digital servos**, and this simplifies the wiring of the arm, since we just plug one servo into the next in a series. Rather than being controlled by analog signals, these servos have a digital serial interface that gives us not just fine control, but the ability to determine where the motors actually are, rather than where they were commanded to be. Why is this important? If the arm hits something and can't continue to move, it will stop. This is not a surprise, but then when you ask for the motor position, it tells you where it stopped. This means you can use the arm itself as a sensor! To determine where the floor is, you command the arm to a downward position, let it stop when it hits the floor, and read the motor positions to see where the floor actually is relative to the arm.

- This arm also lets you move the arm manually – with your hands – and then read the position of the arm, which is very useful when designing poses.

- Another useful feature is the **self-programming mode**. You can put the arm in *program* mode using the relevant button, then move the arm and push the other button (labeled *run*), and the program moves into the arm without the computer. Then you can push *program* again to store these moves, and then hit *run* and it will play back what you did. This is useful for testing out moves. I used it to prototype grasping positions for picking up toys.

If you would like to control a simulated robot arm instead of a real one, there is a tutorial at `https://community.arm.com/arm-research/b/articles/posts/do-you-want-to-build-a-robot`.

Arm controller

There is an important note on wiring up the arm – the arm draws far more electrical power than the USB is capable of supporting. Do *not* run the arm without a power supply connected to the power port. The arm will run with the power straight from the battery, without using the DC/DC converter. I added a master power switch to the robot so I can turn it on and off safely. Without the power switch, the only way to shut down the robot is by physically unplugging the battery, which is problematic.

Arduino microcontroller and motor controller

The Arduino UNO is integrated into the motor controller, which is plugged into the Arduino's **General Purpose Input/Output (GPIO)** ports. This is the simplest way to create a computer interface to the four motors that drive the base of the robot. The motor controller has to have power from the batteries (again without going through the DC/DC converter). The Arduino uses **Pulse Width Modulation**

(**PWM**) to control the four brushed DC motors in the drive base. The motor controller is from Adafruit, and is available at this website: `https://www.adafruit.com/product/1438`.

Sensor – USB camera

The main sensor of Albert is the wide **field-of-view** (**FOV**) camera that has a USB interface. Any number of cameras will fill this need, including many webcams. My camera has a 170-degree FOV and a resolution of 1,024x768 pixels. You can use a camera with higher resolution but make the USB ROS camera driver downsample the image to 1,024x768 so that we know that the rest of the software can handle the bandwidth. I used a color camera with an RGB (three-color) output.

Sensor and effector – audio interface

I bought a USB audio card (`https://www.amazon.com/gp/product/B08R95XJW8`) to support the voice input and output on the robot. This audio interface has both a microphone and some small speakers. This unit makes our interface to the voice interface for speech recognition and text-to-speech output. This unit is well constructed and sturdy and it does a good job with music as well.

Here is a handy wiring diagram of how Albert goes together:

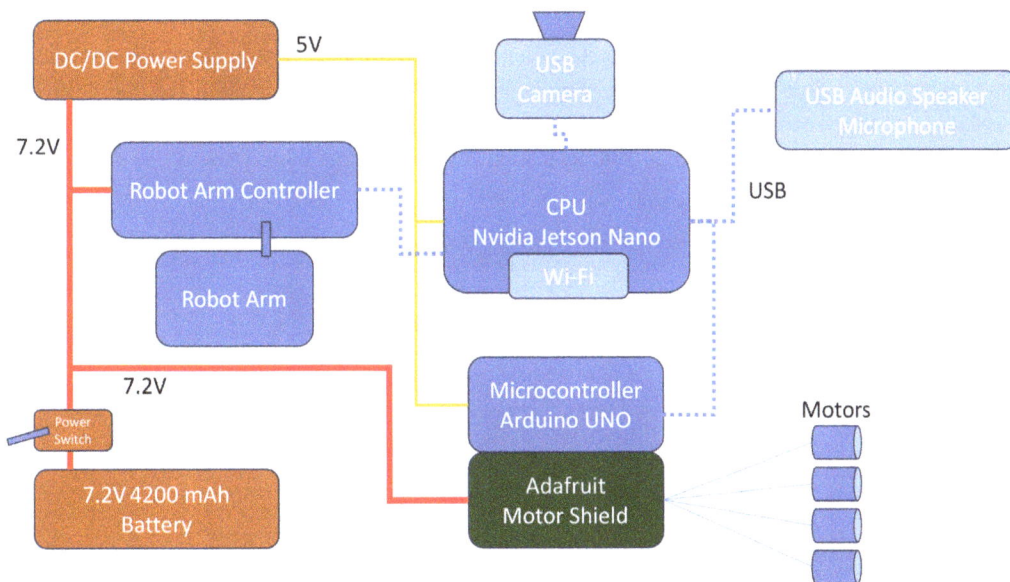

Figure 12.2 – Block wiring diagram of Albert the Robot

Now that we have Albert the Robot assembled, we can use it for the examples in the book. Albert's a very versatile platform with a lot of capability, as you will see.

Robot safety tips

Let's quickly look at some safety tips related to working around robots:

- We are using fairly high-current batteries and drive systems. Be very careful with wiring and look out for shorts, where the positive and negative wires are touching. It is not a bad idea to put a fuse between the battery and the power supply of about 10 Amps to protect against accidental shorts.

- Be careful when the robot is operating. It may suddenly change direction or get stuck. I have a policy of not sitting down, having my hands in my pockets, or using a cell phone when the robot's motor drives are activated. You need to be paying attention.

- Beware of the pinch points in the robot arm – you can get a finger caught in them quite easily (as I have learned). Don't put your fingers inside the joints with the power turned on.

- Have a checklist for setting up the robot, starting all the software, and turning on the hardware. This will stop you worrying that you have forgotten a step.

- In general, here are the steps you should be following. Turn on the robot's power. Wait for the computer to boot up. Connect to the onboard computer from your laptop or desktop using **Virtual Network Computing** (**VNC**). Start the onboard software, and then start giving commands.

- Beware when charging lithium batteries; they can catch fire. It is best to charge them inside a metal box. I used NiCad batteries in my Albert prototype for this reason – no fire hazard. Lithium batteries are lighter and stronger, but they have their downsides as well, such as catching fire, being a hazardous chemical, and permanently losing charge when frozen.

Index

‹packt›

Other Books You May Enjoy

If you enjoyed this book, you may be interested in these other books by Packt:

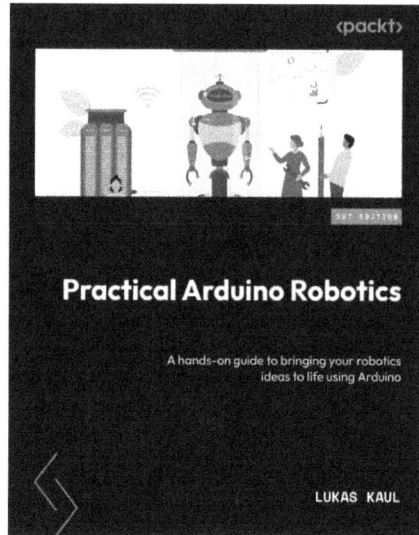

Practical Arduino Robotics

Lukas Kaul

ISBN: 978-1-80461-317-7

- Understand and use the various interfaces of an Arduino board
- Write the code to communicate with your sensors and motors
- Implement and tune methods for sensor signal processing
- Understand and implement state machines that control your robot
- Implement feedback control to create impressive robot capabilities
- Integrate hardware and software components into a reliable robotic system
- Tune, debug, and improve Arduino-based robots systematically

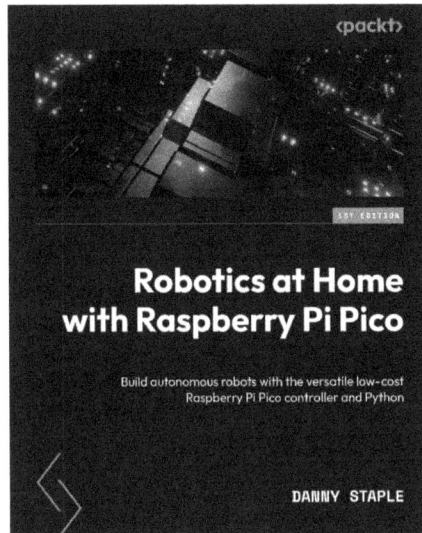

Robotics at Home with Raspberry Pi Pico

Danny Staple

ISBN: 978-1-80324-607-9

- Interface Raspberry Pi Pico with motors to move parts
- Design in 3D CAD with Free CAD
- Build a simple robot and extend it for more complex projects
- Interface Raspberry Pi Pico with sensors and Bluetooth BLE
- Visualize robot data with Matplotlib
- Gain an understanding of robotics algorithms on Pico for smart behavior

Packt is searching for authors like you

If you're interested in becoming an author for Packt, please visit `authors.packtpub.com` and apply today. We have worked with thousands of developers and tech professionals, just like you, to help them share their insight with the global tech community. You can make a general application, apply for a specific hot topic that we are recruiting an author for, or submit your own idea.

Share Your Thoughts

Now you've finished *Artificial Intelligence for Robotics*, we'd love to hear your thoughts! Scan the QR code below to go straight to the Amazon review page for this book and share your feedback or leave a review on the site that you purchased it from.

`https://packt.link/r/1805129597`

Your review is important to us and the tech community and will help us make sure we're delivering excellent quality content.

Download a free PDF copy of this book

Thanks for purchasing this book!

Do you like to read on the go but are unable to carry your print books everywhere?

Is your eBook purchase not compatible with the device of your choice?

Don't worry, now with every Packt book you get a DRM-free PDF version of that book at no cost.

Read anywhere, any place, on any device. Search, copy, and paste code from your favorite technical books directly into your application.

The perks don't stop there, you can get exclusive access to discounts, newsletters, and great free content in your inbox daily

Follow these simple steps to get the benefits:

1. Scan the QR code or visit the link below

https://packt.link/free-ebook/9781805129592

2. Submit your proof of purchase
3. That's it! We'll send your free PDF and other benefits to your email directly

www.ingramcontent.com/pod-product-compliance
Lightning Source LLC
Chambersburg PA
CBHW081049220326

41598CB00038B/7038